ETHICS
&
SCIENCE

Ethics & Science

Henry Margenau

D. VAN NOSTRAND COMPANY, INC.
PRINCETON, NEW JERSEY   TORONTO NEW YORK LONDON

D. VAN NOSTRAND COMPANY, INC.
120 Alexander St., Princeton, New Jersey (*Principal Office*)
24 West 40th Street, New York 18, New York

D. VAN NOSTRAND COMPANY, LTD.
358, Kensington High Street, London, W.14, England

D. VAN NOSTRAND COMPANY (Canada) LTD.
25 Hollinger Road, Toronto 16, Canada
Copyright © 1964, by D. VAN NOSTRAND COMPANY, INC.

Published simultaneously in Canada by
D. Van Nostrand Company (Canada) Ltd.

No reproduction in any form of this book, in whole or in part (except for brief quotation in critical articles or reviews), may be made without written authorization from the publishers.

Printed in the United States of America

Preface

Ethics is a matter of concern to us largely because of its failures; science is the object of admiration because of its successes. But ethics, too, has a working methodology which is arresting in view of its successes and deserves the same sort of painstaking positive analysis science has been getting. Now it seems to me that most accounts of ethics dwell upon its *un*solved problems, upon the reasons why it is not and cannot be as authoritative, as cogent in its appeals to men as is science; upon its lack of uniformity, its so-called relativity; upon the conflicts it is unable to resolve; upon its helplessness without religion. Rarely is there an acknowledgment today of the patent fact that human societies would be impossible except for the order and uniformities ethics has indeed achieved, or of the usefulness of ethics as that singular accomplishment which elevates man above nature and thereby enables him to develop such disciplines as science. The present analysis takes cognizance of ethics as a successful human enterprise and parenthetically notes its failures.

I detect in the writings of contemporary moral philosophers another trait which I deplore. Ethics may be likened to a vehicle transporting passengers to a mountain top. As it moves, it reflects an occasional gleam into the eyes of distant and interested observers, who are fascinated by these gleams and call them values. And of such brilliant reflections moral philosophers attempt to construct the appearance of the vehicle and to understand its action. The effort here is to go up to the vehicle, to study its parts and its structure, to see what makes it move and to understand in the end, not only why it gleams but also why it works.

Finally, I wish to record here my vehement reaction against a point of view which was so forcefully expressed by Nietzsche (*The Twilight of Idols and The Antichrist*) when he said:

> THE GOOD-IN-ITSELF, THE GOOD STAMPED WITH THE CHARACTER OF IMPERSONALITY AND UNIVERSAL VALIDITY—THESE ARE MERE DELUSIONS IN WHICH THE DECLINE, THE ULTIMATE ENFEEBLEMENT OF LIFE . . . EXPRESS THEMSELVES. THE MOST FUNDAMENTAL LAWS OF PRESERVATION AND GROWTH DEMAND PRECISELY THE OPPOSITE, NAMELY THAT EVERY INDIVIDUAL INVENT HIS OWN VIRTUE, HIS OWN CATEGORICAL IMPERATIVE.

This is a vicious anachronism. It is an anachronism because the knowledge on which it is based is drawn from a science that featured mechanical causation and the tooth-and-claw doctrine of life, a science which has been abandoned. And it is vicious in our day because the survival of mankind depends upon the communality of ethical standards, upon the communality of virtue. Not only do we need transindividual criteria of goodness, we need more than ever transcultural standards by which the ethics of nations can be judged. I cherish the hope that the sketch drawn in these pages, which exhibits parallels between ethics and science without committing the "naturalistic fallacy" of reducing ethics to science, will confer upon moral philosophy a measure of that universal authority and attractiveness now enjoyed by science.

The writing of this book was greatly aided by a grant from the Rebekah M. Harkness Foundation, administered through the good offices of the Foundation for Integrative Education, whose president Kirtley Mather and executive officer Fritz Kuntz gave their wholesome encouragement to this venture. For the trouble they took in reading early drafts of the manuscript and for their helpful comments I am indebted to the following friends and colleagues:

H. G. CASSIDY,   C. J. DUCASSE,   E. HARRIS
E. H. MADDEN, F. S. C. NORTHROP, N. M. SMITH

My major debt of gratitude is to my wife who, while still skeptical about the need of books like this, has enriched my life by innumerable practical examples of the meaning of "good."

HENRY MARGENAU

Contents

CHAPTER

		PAGE
I	THE METHOD OF SCIENCE	1
	The Popular View of Science	3
	Science and Human Experience	9
	Postulates and the Structure of Scientific Theory	30
	Induction and Deduction	41
	Verification	44
II	THE METHOD OF PHYSICAL SCIENCE AND HUMAN AFFAIRS	54
	The Problem	55
	Six Arguments Purporting to Prove that Scientific Method is Inapplicable to Human Affairs	62
	Physical Laws and Social Behavior	74
	Second Thoughts on the Six Arguments	83
III	RÉSUMÉ ON VALUES	98
	Preliminaries	99
	The Meanings of Value	104
	Methods for Assessing Value	113

	The Scale of Oughts	120
	Does Science Contain Normative Elements?	125
	Theories of Value	129
IV	THE WORKING METHODOLOGY OF ETHICS	138
	Statement of Aims	139
	"Axioms" of Ethics	144
	Explication of Axioms	158
	The Protocol Plane of Ethical Experience	164
	The Validation of Norms	168
V	NORTHROP'S ETHICS	181
VI	THE HISTORICAL CONFUSION BETWEEN VALIDATING PRINCIPLES AND ETHICAL IMPERATIVES	199
	The Postulational Character of Primary Values	200
	Theories of Moral Emotions	214
	The Connection between Primary Values and Commands	221
	The Intuitive Connection between Commands and Primary Values	230
VII	ORIGINS OF IMPERATIVES AND PRIMARY VALUES	233
	The Induction of Ethical Attitudes by Non-Ethical Doctrines	234

	The Anomaly of Christian Ethics: Paul	241
	Ethics and Divine Mercy: Luther	249
	Ethical Systems which are Loosely Coupled with Metaphysical Views	253
	Summary and Conclusions	255
VIII	CONSEQUENCES OF THE PARALLELISM BETWEEN ETHICS AND SCIENCE	260
	A Second Look at Values	261
	Ethical Conflict	266
	Conscience	270
	Ethics as a Group Enterprise	276
	Ethical Relativity	281
	Obligation vs. Purpose	288
	Conclusion	291
AUTHOR INDEX		295
SUBJECT INDEX		299

I

The Method of Science

ABSTRACT

When certain errors in the popular view of science are corrected, science is seen to be applicable not only to facts and quantities but also to many subtler essences present in human experience. To make this clear, experience in its fullness is analyzed into components called cognitive, appreciative, voluntary, and others. Natural science organizes cognitive experience because it is deliberately designed to seek that end; its resignation to a limited purpose must not without further thought be taken as an intrinsic incompetence.

An analysis of cognitive experience reveals factual and rational ingredients together with a third class, called rules of correspondence or epistemic correspondences, which mediate between these two. The genesis of concepts proceeds simultaneously with the discovery of perceptory facts, and the latter serve as protocols for all rational developments. Usually, science performs these tasks in systematic stepwise fashion, but occasionally a blessed individual is able to perform an inductive leap from a bewildering array of facts to a harmonizing theory. Psychologically, such acts are akin to inspiration and revelation in ethics and religion.

As scientific theory moves further and further away from the protocol domain of observable facts, it tends to become increasingly abstract. This movement is limited; the elements forming the limit are called axioms or postulates. Their truth cannot be demonstrated directly but only through their empirical consequences, which never exhaust the meaning of the postulates. Hence the postulates are never fixed; they impart to science that welcome flexibility in the face of new facts which make it a progressive, self-corrective enterprise.

THE METHOD OF SCIENCE

THE POPULAR VIEW OF SCIENCE

In speaking of "the world of gadgets called science"[1] John Steinbeck doubtless expressed a popular point of view. Practical expediency is the watchword of science, robots without souls are its products. This is because science has a keen eye for facts, limits itself to facts and by this limitation achieves the tremendous mechanical successes which enhance and at the same time threaten our lives. One of the tricks science employs is measurement, the elaborate establishment of numerical values; and through its quantitative procedures it gains sinister efficiency, but it also squeezes the essence out of things; in particular it fails to account for values. The world, after all, presents qualities as well as quantities, and in neglecting the former, science is able to deal more adequately with the latter. By the same token, however, it is forced to resign before those subtle matters that form the substance of man's ethical and spiritual being. Such, briefly stated, is the thesis against which the arguments in the first chapter of this book are directed.

Let us call it the popular view of science. It is erroneous in two respects: first, because it makes several mistakes in fact, and secondly, because it is grotesquely incomplete, ignoring precisely those parts of science which in the opinion of its practitioners form its living substance. Among the incidental errors is the belief, expressed above, that observa-

[1] *Esquire,* "A Primer on the 30's" June 1960, pp. 85-93.

tion in the form of measurement commits science to *quantitative, numerical judgments,* and that such judgments are in some way limited. As to measurement, what it is and does will be discussed in some detail later in this chapter. In general it is wrong to say that measurement furnishes only numbers, as is the custom in older treatises on the philosophy of science. In the scientifically important sense measurement often decides questions of existence; the spectroscopist looks for the occurrence of a certain line on his plate; the nuclear physicist, for a track in a cloudchamber. Such existential observations are even more crucial, more primitive than the acts called measurement, and they abound in science. As to measurements themselves, they do yield numbers *by definition.* But to regard this as a limiting or degrading circumstance is quite unwarranted, for the statement is both true and trivial and does not fully characterize the meaning of measurement. It is as true and as trivial as the statement that a painting consists of brush strokes, or that a fine watch consists of molecules, which misses the important things about a watch, the intricate wheels and springs and gears into which the molecules have been fashioned, the delicate positioning of these mechanisms that makes the whole thing work. And so in science: measurements are its basic ingredients, but judicious *combinations* of measurements produce not *mere* numbers but functions, fields, matrices, groups and many other theoretical entities whose nature is very poorly described by saying they are numbers based on observations.[2]

Are they then truly *quantities,* to which the popular view restricts the competence of science? A mathematical group represents certain properties of crystals. In a strict sense these are non-numerical. They portray symmetries, the very essence of beauty in ornamental design. Are symmetries quantities, or are they qualities? If the former, then this kind of beauty has been quantified, and esthetics has been "reduced" to science. If the latter, the popular view of science is wrong. Without pursuing this matter further, let us draw here the minimal conclusion: that a clear and hard distinction between quantities and qualities is difficult to establish, that the elaborate procedures of every advanced science lead to stages where it cannot be maintained and that in the history of science many an elusive quality, regarded in the popular view as the quintessence of the non-scientific (e.g., certain kinds of value) has yielded to the scientific approach. And, strange to say, it survived the "kiss of death" and flourished afterwards. Science is continually evolving new and better quantitative facilities for dealing with the qualitative aspects of things.

The popular view commits another error when it identifies science exclusively or even predominantly with *discovery of facts.* To be sure, there is a good deal of conven-

---

[2] For further discussion of such non-numerical elements in modern science see pages 64–65, 86 et seq.

tional truth in this characterization, but the truth is due almost entirely to the looseness of the word fact. Science is certainly full of conjectures and ideas, some of them bolder than anything one might encounter outside of science. Its visions are keen and lofty, they may be flights of fancy differing from poetry and fiction only in that they allow themselves to be confirmed by careful observation. Would anyone apply the word fact to gravitation, that unimaginable influence which stretches between bodies of interstellar space, inaccessible to vision yet highly structured and precise in its mathematical form? What about the libido and the subconscious of psychiatry, unified field theories, the time contraction of relativity—are they facts? They are, I suppose, if "being a fact" is merely one of the many meanings of "true." This, however, is hardly what the popular view suggests: it wants to tie science to the obvious stuff we find about us, to what the senses convey; it wants to commit science to the reportorial role of an accountant who keeps track of what he sees, hears, feels, tastes, and smells, as Ranke committed history. But in this insistence the popular view is completely wrong.[3]

For while science does and doubtless always will make contact with the external data impinging upon us, it tran-

---

[3] For further elaboration of this point see H. G. Cassidy's memorable *The Sciences and the Arts: A New Alliance*, New York: Harper, 1962; the author's *Open Vistas*, New Haven: Yale University Press, 1961; and the excellent recent book by R. B. Lindsay, *The Role of Science in Civilization*, New York: Harper & Row, 1963.

scends them by far. It certainly cannot get along without logical and mathematical procedures. Everybody knows this, of course, but the tendency to forget it or to minimize its importance is all too strong. Thence results a strange popular misconception of the term *theory,* which is supposed to be the antithesis to fact. When a newspaper article labels an explanation a theory it means to damn it as uncertain and probably erroneous. Contrast this usage with the employment of the word theory in the literature of physics, where it puts the last stroke of perfection upon our understanding of phenomena previously obscure and enigmatic in their factualness. Many would go so far as to say that theories are the dynamos of science, its generators of power and, in the last analysis, of its practical usefulness. Einstein's simple theoretical formula, $E = mc^2$, has had consequences that threaten to destroy mankind. Let us not underrate scientific theories.

While theories differ from facts, another element vitally involved in every mature science differs still more radically. This is the material which forms the logical starting point of science, its axioms. Deductive science (in contradistinction to descriptive sciences like geography or botany) begins with fundamental, *unproved* propositions which are verified only in their several consequences. The scientist does not seek to prove these axioms; rather, he accepts them provisionally, judiciously but without proof, hoping that their *consequences* agree with the facts. Nor is his attitude toward them one of avoidance or tolerance. He cannot get going without them in spite of the circumstance that they represent precisely the opposite of what the popular view takes to be a fact.

The foregoing remarks voice criticisms of an interpretation of science that seems to be widely held. If the reader feels that they are arrows shot at a straw man set up as an artificial target, I ask his indulgence. It would be a relief to know that these corrective strictures are unnecessary. In the following pages of this chapter an attempt will be made to outline the structure of science in positive terms, to summarize the large distinctive features of scientific inquiry that cause it to differ from pursuits which do not lead to clear and certain knowledge. Philosophers would speak of what follows as an exercise in epistemology, the theory of knowledge (Greek: *episteme,* knowledge; Greek: *logos,* theory) developed with particular reference to science. In the full sense of that word common sense, too, is part of science; it is the residue left in the wake of advancing science which then diffuses into popular understanding and becomes generally accepted; cosmology and other speculative disciplines are likewise sciences in the wide sense here employed. What is to be sketched, then, is not some narrow methodology portraying the procedures in a few special sciences but a very general approach available in all fields where certain knowledge has been or can be established. A narrower frame would exclude ethics.

At this point a brief note on the semantics of the word

THE METHOD OF SCIENCE

"science" should be added. The universal meaning of its ancestor, *scientia,* has not survived in English, although it is preserved in French and in its German descendant, *Wissenschaft* which implies accurate scholarly knowledge in all fields, including the *Geisteswissenschaften.* The word science has reference to such large areas of investigation, physical and social science. It is generally agreed that, because of the spectacular successes of the physical sciences, these represent the native and primary habitat of scientific method; its applicability to social science is not generally accepted. Whatever the case for social science may be—and we shall deal with it in Chapter II—our present concerns are centered in the physical sciences, and our study of method proceeds from there. Nevertheless, it is hoped that the treatment presented can be seen to involve nothing which, on the face of it, precludes the aims of social science.

SCIENCE AND HUMAN EXPERIENCE

As a first step the meaning of *experience* must be analyzed. The Latin verb *experiri* adverted to all possible phases of awareness, to perceptions, visions, hallucinations, feelings, thoughts, judgments, decision, actions, and innumerable other unnamed conscious processes man can undergo. Mainly through the teachings of British empiricists (Locke, Hume and their dissenting disciple, Kant), the range of denotation of the word contracted until it came to designate only the activities which are involved in or immediately consequent upon external perception. Some see in this transformation a perversion of language—be that as it may: the claim that this narrow sense of the word contains all that is vital to *knowledge,* a claim made by extreme empiricists, is certainly a perversion of philosophy. Hence we return here to the original meaning of experience, allowing it to include what has been called the affective, the emotional, the conative, and all the rest.

One might ask whether the use of so embracing a term will not result in uninteresting and pointless statements. This is not the case, for the term retains exactly that discriminating power which is so important in philosophy. It allows the distinction between what is within experience and what is outside it, giving occasion for example to the meaningful inquiry concerning matters that transcend experience, making possible the time-honored distinction between epistemology (which remains within experience) and ontology (which may transcend it). We do not intend to ask such questions here but wish to record their validity, even while restoring this very extensive range of meaning to the term experience.

But now we trim it down; we select from the large domain a certain class of experiences called *cognitive*. Roughly, they are the components which lead to knowledge or understanding as distinct from purely affective moods, appreciative or depreciative attitudes, decisions and actions. These terms are admittedly vague, and there is nothing we can do about it. Experience does not come in neatly classified compartments; it is a stream with certain accentuations which we recognize and which we label feelings, thoughts, etc., much in the way we call a color blue. None of these elements of experience is strictly definable; each is a primitive in a logical system one might wish to base on them. Bearing this in mind, we characterize cognitive experience by listing the words commonly employed to denote its fluid components, naming sometimes the psychological activity, sometimes its results in mind (since both are experiences, one active, the other contemplative): Sensory awareness, perception, external data, observation, abstraction, judgment, analysis, thought, concept, reasoning, inference, conclusion, induction, assumption, prediction, verification, and many more, including that synthesizing bond, which makes for unity of experience, namely, the *memory* of previous items of awareness.

Traditionally, these are often divided into two large groups: perceptions yielding external data on the one hand and reasoned judgments yielding concepts on the other. Memory, though not ordinarily included, must be considered to hover above the scheme, ready to be called upon when needed. Different motives have induced different philosophers to make the distinction between data and concepts; some intended to convey the origin of the experiences, assigning data to external objects and concepts to men's minds; others wanted to distinguish their reliability, data being incontrovertible while concepts are derivative and subject to error. The reverse position as to order of reliability was held by some rationalists. A third motive for wishing to distinguish between data and concepts is one which has seemed most significant to the present writer;[4] every perception of a datum bears clear testimony to its spontaneity and coerciveness. Data inflict themselves upon us, whereas concepts are distinctively of our own making. We have, we receive, we accept, we suffer the data which assail us through our senses; we abstract, select, combine, invent, create the concepts with which we reason. There is a sense of fatefulness about the acceptance of what perception and observation deliver, a sense of responsibility about the results of reason. In the latter instance, we feel ourselves involved like an architect in the structure of a building; in the former, somewhat as an agent who acknowledges the receipt of building material.

The difference just sketched is incorporated into the method of science. Because the scientist has a right to disclaim responsibility with respect to data, he may treat them as independent of his own procedures, as "given"; he may look upon them as something ultimate in his experience which he needs to accommodate and assimilate in other phases of his experience. Figuratively speaking, datal or per-

---

[4] See H. Margenau, "Methodology of Modern Physics," *Philosophy of Science*, Vol. 2, No. 1, January, 1935; *The Nature of Physical Reality*, New York: McGraw-Hill, 1950.

ceptory experience is a crude suggestive forerunner of more elaborate and rational kinds of cognitive experience, much in the way in which the *protocol* of an ancient book, i.e., the leaf pasted inside the cover carrying the verbal plan of the book, forecast its contents. I have borrowed the word protocol from Carnap, whose early writings use it in a similar sense. The kind of experience which, roughly speaking, is not of our making will here be called P-experience. The reader may think of P as standing for protocol, perceptory or primary. In science it conveys a kind of irreducibility, of authority, which concepts lack.

Whether sense data and, more generally, observation based on and involving perception are alone in having this P-quality is a question we need not answer in this book. It is my belief that there are other experiences which can claim this property, cognitive experiences which are as coercive, spontaneous and reliable as perceptory data. They include perhaps the records of history, the introspective data of psychology, the immediacies of religious experience, possibly even paranormal perceptions. Natural science, however, has not used them as such, and our analysis is here limited to the domain of natural science.

To signify the components of cognitive experience which stand at the other extreme of the range, far away from P, we shall use the letter C. Again, it may be regarded as an abbreviation of "concept." But this word has unfortunately become the bearer of terrific burdens. In philosophy alone it seems to designate anything that can be thought, from the image of a particular object presented in sensation to the abstract universal "being." Add to this the meanings of our military writers who are displaying an increasing fondness of the word, and its suitability may become questionable. For these reasons, and also because a reference is needed to the distinctive fact that the concepts are of our making, the word construct has been employed[5] elsewhere, and C may be taken to allude to it. The danger here, of course, is that the word might be mistaken for "mere, subjective, mental" construct; such adjectives do describe the usual primitive psychological situation in which constructs first arise, but the possibility of confirmation by comparison with P-experiences lifts them out of this primitive subjectivism and converts constructs into "verifacts," to use once more an earlier terminology. F. S. C. Northrop[6] speaks of concepts by postulation and concepts by intuition; these are approximately the present writer's C and P experiences.[7]

The relation of these epistemological factors and "reality," while of secondary importance for the problems of ethics, is sufficiently interesting to warrant comment here.

In the *Nature of Physical Reality*, the author, in search

---

[5] H. Margenau, *loc. cit.*

[6] F. S. C. Northrop, *The Meeting of East and West*, New York: Macmillan, 1947; *Logic of Science and Humanities*, Stillwater, Oklahoma: Oklahoma A & M College, 1950.

[7] F. S. C. Northrop, *The Nature of Concepts, Their Interrelation and Role in Social Structure*, Stillwater, Oklahoma: Oklahoma A & M College, 1950.

of the clearest statement of what scientists mean by the term "physical reality," defined it (roughly) as the set of all verifacts and all P-experiences. Other possible definitions of reality were surveyed in the last chapter of that book. Physical reality, according to this definition, changes as knowledge changes; and this feature, admittedly unsatisfactory to common sense realism, has given rise to apprehension and criticisms, not of the method presented but of the definition given the term, physical reality, which was catapulted into prominence because it appeared in the title of the book.

The critic whom I most respect is my old friend and colleague Professor Filmer S. C. Northrop, who more than anyone else stimulated my interest in philosophy and with whose views I am very largely in sympathy. His unique contribution to the Credo Series entitled *Man, Nature and God, a Quest for Life's Meaning,* is a book of profound insights, informal and sometimes poetic excursions interspersed with serious passages devoted to "straightening out" the philosophic record, a book of infinite charm. In it he honors me by taking issue with my views, an honor that would be ill bestowed if I did not reply in an effort to stand my ground. This place seems opportune for such defensive comment; hence the next few pages will be devoted to a discussion of Professor Northrop's strictures of the thesis outlined in the preceding paragraphs.

He says: according to this writer's view "nature . . . became, so far as any realistic objective reference is concerned, a mere symbol-projective *als ob,* as the post-Kantian Vaihinger saw." This passage seems occasioned by a misunderstanding of what I endeavored to say. If "nature" comprises physical reality—and this is what most scientists take it to mean[8]—it refers to P-experiences and verifacts. These, in the first place, are not symbols of anything; they are themselves the nodal parts of human cognitive experience and stand for nothing else. Secondly, they have no reference to Vaihinger's fictions, which were imperfect models of some unknowable, transexperiential reality. Nature, as I conceive it, is what we experience, and if what we experience is in some vague Kant-Vaihinger sense a fictive as-if version of some unknowable transcendental (in pre-Kantian understanding) reality, then the latter is certainly not to be regarded as nature, while the former must be so identified. Thirdly, Professor Northrop's sentence, perhaps inadvertently, has nature, i.e., the verifacts and the immediate sensations, *project* symbols. The best one can do here is to suppose that I project them; but this interpretation fails for the two previous reasons since there are no symbols to be projected: There is only experience to be described and organized.

In another place Professor Northrop suggests: "(Invari-

[8] In some parts of *The Nature of Physical Reality,* the capitalized version, Nature, was used to designate P-experiences alone in an effort to exploit the literal, personalized, etymological meaning of the word as birthgiver, as that which emerges and gives rise to emergence, in the very sense of Professor Northrop's "perishing particulars." I do not choose this meaning here.

ance) did not, in Cassier's mind, nor does it with Margenau, warrant the thesis that concepts by intellection are to be interpreted logically realistically as designating external events and objects in ontologically existing nature."

If I understand this statement correctly it is true, but true mainly because of the complexity of its qualifications. If the thesis in question were simply: "Concepts by intellection designate external events and objects" I would certainly affirm it, provided the concepts by intellection are taken to be verified constructs, verifacts. Indeed, *The Nature of Physical Reality,* which is the target of Professor Northrop's critique, is an exposition of the rules which entitle us to the claim that constructs designate external events and objects. It should be added, however, that the analysis conducted remained deliberately within the framework of epistemology and therefore was forced to lodge external events and objects within experience. Professor Northrop wishes to place them "in ontologically existing nature," as common sense would persuade us to do.

The phrase, ontologically existing nature, troubles me. I know what existence means, at least in the sense in which physical objects exist. It is the status of the desk on which I write and, in a different mode, of myself. These are quintessences, distillates from very common phases of my experience, forced upon me by certain invariant features which it exhibits. And from further factors within my experience, which includes other persons, I infer by the very principles outlined in *The Nature of Physical Reality* that these objects which exist for me exist equally for other persons. But Professor Northrop requires that they also exist "ontologically."

Even here I would follow him to a certain point. If ontological[9] means "apart from being known" then the reality I postulate of, or ascribe to, any object entails its ontological existence. If it means existing prior to any knower I am still in sympathy with the assertion that things exist ontologically; that was actually shown to be a consequence of the scientific method (see *The Nature of Physical Reality*).

Unfortunately, however, Professor Northrop requires more than this. He interprets ontological as "the same for all knowers," thereby pushing the assertion not only beyond verifiability but also past the point of credibility. For it seems clear that the real desk before me, which Eddington conceived as a myriad of electrons whirling about nuclei, is different to the person who is ignorant of modern physics; that an artist's conception of the evening star is different from an astronomer's. More important even is the fact that real things are different to knowers at different epochs of science, or even to the same knower in different stages of enlightenment. Thirty-five years ago I "knew" that the atomic nucleus consisted of protons and electrons; twenty years ago I "knew" it to be composed of protons and neu-

---

[9] In languages like English and German, which have only one verb, to be (exist is an almost exact synonym), that verb groans under a load of different and uninspected meanings. Orientals are more discriminating, and I have found it interesting that "ontology" means very little to them.

trons; now it is infested by a variety of unstable entities, and furthermore there has evolved the practical certainty that the present picture will change as research progresses. This is what prevents us from accepting the dictum: reality is the same for all knowers.

It might be objected that the verb, to know, was used illicitly in the last two sentences, that precisely because our understanding of nuclear structure was bound to change, I did not know. This, however, dissolves scientific knowledge, the most certain knowledge we have about nature, to naught and deprives the verb, to know, of every shred of significance. The last sentence of the previous paragraph must therefore be allowed to stand.

Lest I be accused of charging Professor Northrop with carelessness I hasten to acknowledge that his writings bear every evidence of his being aware of the difficulty just cited. Surely, the phrase in question, "the same for all knowers," could not be meant literally even though it appears in very crucial contexts on pages 218, 221, 224 of *Man, Nature and God*. Indeed there is the simple and unmistakable brief sentence: "Real means every item of knowledge that is the same for all knowers." As we have seen, this limits reality to the most trivial experiences or facts, for instance the observation: it is now 9:05 P.M. in New Haven—and even here one might doubt the sameness to all knowers.

My reason for assuming that Professor Northrop knows all this, aside from my high regard for his wisdom and his philosophic circumspection, is found on another page in his remarkable book. For he himself voices misgivings about the meaning of existence, or at least its present definability, on the basis of his logical realistic position when on page 224 *et seq.*, he holds that his "thesis has to be stated with great care. The word necessary to provide this care is 'asymptotic.'" What is the same to all knowers, we are then told, is the *limit* which human knowledge asymptotically approaches. There is the suggestion, not explicitly stated but clearly necessary to the argument, that except for this limit the asymptotic convergence of knowledge, the progressive verification of constructs, is meaningless and that *therefore the limit must exist.*

This is a *petitio principii.* The method of reasoning here is borrowed from mathematics. There, however, this argument would not work, for one can construct innumerable ever progressing series which do *not* have limits. Mere progression is never a guarantee for the existence of a limit; least of all does it *define* the limit, even if that limit exists, in a manner that makes it known to all, and certainly not "the same for all knowers."

Nevertheless I believe this argument has very great psychological force; the hope, often amounting to a conviction, that knowledge approaches a knowable, ideal limit is one of the strongest motives in scientific research. This view was favorably discussed in *The Nature of Physical Reality,* where it was listed as yielding a possible definition of reality. I did not identify it as *physical* reality, to which that book was devoted (nowhere did I suggest that physical reality is the only type of interest to the philosopher!), but observed that the view was a plausible combination of two theses: a) the avowal of the methodology worked out in that book, and b) the added postulate or belief that human knowledge converges. The latter, it was pointed out, is not part of a philosophy of science, nor of epistemology proper, but belongs to the philosophy of history. On this score, then, I take pleasure in observing that Professor Northrop's view does not differ essentially from my own.

Strange, however, is the name he gives to his position when he calls it logical realism. Asymptotic realism would seem understandable, but why that which does not follow logically from any proposition known to man but evolves slowly in scientific procedures, in case it does evolve, should be called "logical" escapes this author's comprehension.

The patent fact is that Professor Northrop is more of a Kantian than the neo-Kantians (among whom he includes Poincaré, Cassirer and myself) he criticizes: He wants the *Ding-an-sich* while I, for one, am willing to get along without it so far as science is concerned. I avoid it, for instance, in defining *physical* reality. As suggested in *The Nature of Physical Reality,* to pass from physical reality to other kinds of existence one needs the guidance of principles beyond those discussed in that book, principles I hope to consider in a volume under preparation.

On page 222 of *Man, Nature and God* even Einstein is called a logical realist. Whether he would have accepted that classification is beside the point; but what is said in that context appears to be against it. For he is quoted as identifying "the axiomatic substructure of physics" with "our conception of the structure of reality." This defines precisely what I called physical reality, the kind that changes irreversibly (i.e., when changes in the axiomatic substructure occur, these changes are retroactively injected into the interpretation of past reality) as the axioms change and is therefore not the same for all knowers.

Professor Northrop's strongest argument against my view is marshalled on page 226, where he states that a person holding the "nonrealistic interpretation of concepts by intellection . . . must believe that his wife, or anyone else dear to him, has no ontologically realistic objective meaning . . ." If those whom I love are mere verifacts in my personal experience my affection is somehow deprived of its substantial object; it is reduced to an epistemological relation meriting analysis but not sentiment, and love subsides into mere intellectual interest. It is exactly because of these conclusions that I deem physical reality, and for that matter science, insufficient to sustain the living qualities of human existence. This, however, does not invalidate the point of my definition of *physical* reality; for it is primarily because my loved ones exist for me not only as physically real but as persons having qualities which transcend this elementary mode of being, that

I hold them in affection. Professor Northrop's suggestion, namely that I regard them as logically real in his asymptotic sense, would hardly occasion or justify my love.

Let us return to the more straightforward problems in philosophy of science, which are much less debatable than the nature of physical reality. Nowadays, when discussing questions of this kind, one is often asked, "where does language fit into this scheme?" Some philosophers believe the suggested classification of cognitive experience into P and C domains to be merely a matter of language. To see that this is not the case, and also to observe more clearly the relation of language to experience, one must first realize that language, itself, is a form of experience, sometimes cognitive and sometimes not. One may indeed distinguish between direct (or non-symbolic) and symbolic experience, the latter including speech. There are of course other symbolic forms,[10] such as the primitive grunt, the gesture of pointing, the various art forms, mathematics and logic. Every communication involves symbolism, but not every symbol is for the purpose of communication; uncommunicated private experience can be symbolic: a seen rose can elicit the private sentiment of love. More important for our purpose is the fact that there are experiences which are never communicated, never symbolized, and therefore never expressed linguistically. To be sure, this may be because they are trivial, and

[10] E. Cassirer, *The Philosophy of Symbolic Forms,* New Haven: Yale University Press, 1953.

the subject does not care to communicate them; but it may also be because they are *ineffable,* are of a type for which symbols of communication do not exist.

It is true that most ineffable experiences, especially when they recur, clamor for symbolic formulation, usually for expression in language. And this drive normally succeeds, though not with equal success in all instances. Enrichment of language is the result. It is said that the mystic, upon attaining to ecstasy, has intrinsically ineffable experiences. This statement has one or both of the following meanings. The mystic has a protocol experience which others are strictly incapable of having, like the experience of red to the color blind. Or the experience is so unique, even to the mystic, that he has no words for it, yet he believes he can encounter it again and can induce others have it. Only the latter meaning is of interest in science (and in ethics). Ineffability thus results from the temporary insufficiency of symbolic forms; it attaches primarily to new and unaccustomed modes of experience and is not confined to the P domain. Novel mathematical insights are as difficult to express as the subtle shade of an unusual color or the feeling of delight at special beauty. Man has been called a symbolic animal; he lives under the compulsion of transforming all his experience into symbolic forms.

The difference between P and C is not based on language; indeed language frequently obscures the difference. To note this is especially important for a fuller understanding of the method of science, and a few examples will now be introduced to illuminate this point.

The word force may mean a P-experience, something immediately apprehended as a muscular sensation, as a push or a pull. In the physicist's language force has a different connotation; it refers to something rather more abstract, existing apart from tactile and kinesthetic sensations yet related to them in a unique and measurable way. Newton's law speaks of it as mass times acceleration and implies an entity clearly different from sensed pushes and pulls. We are facing here the C-aspect of force, a new component of scientific experience for whose genesis *we* are responsible, for it is not given as a simple perceptory presentation. There is no reason to suppose that Leibnitz' feeling of the muscular sensation called force was different from that of Descartes, yet they differed in defining the construct force, Leibnitz taking it to be mass times velocity squared, Descartes as mass times velocity, both disagreeing with Newton. One of the purposes served by the C meaning is to connect and regularize seemingly discordant protocol items, to allow measurement, to quantify in a more reproducible manner than direct sensation permits; another is to make the term more "objective," whatever this may mean. Indeed ordinary language often speaks of P as the subjective, of C as the objective aspect of force. Precisely how the two are related, and what the word objective implies, will be considered later. Let us for the moment continue to look at the contrast between P and C in a few other physical quantities, every one of which, as we shall see, presents the dual aspect without its being recognized in our language.

Temperature to the non-physicist means hotness or coldness, something given in immediate sensation. It goes without saying that this sensation is not what is to be inserted in the equations of thermodynamics, nor what is measured by thermometers. To be sure, when reading a thermometer, visual sensations take the place of the sensed temperature which was felt when a finger was inserted in the hot liquid. Yet the visual sensation of the scale mark which coincided with the top of the mercury column, was not identical with the construct, temperature, either. C-temperature arises in a complex way out of a specific group of P-experiences *plus certain inferences* from them; P-temperature, on the other hand, is the direct sensation of hotness.

One more example may suffice. The P-aspect of color is the seen hue in its sensory vividness, the C-component is the frequency of an electromagnetic wave.

It will now be clear that all words denoting measurable physical quantities like sound, noise, pitch, energy, momentum, length, time, mass, etc., have this dual reference, once to protocol experiences and once to scientific concepts.

Earlier in this chapter we have noted the confusion one often encounters in treatments of the quality-quantity distinction. That polarity *can* be maintained in meaningful fashion if qualities are taken to be P-experiences in their esthetic, pure immediacy, while quantities are the constructs associated with them. Thus it makes little sense to say, "colors and sounds are qualities," unless one specifies "directly perceived colors and sounds." Colors and sounds, conceived in the physicist's manner (as electromagnetic waves and material vibrations) are quantities. The quality-quantity relation is then seen to coincide with the relation between the subjective and the objective.

How are the two aspects related? In the instances cited they are related by instrumental (or operational) definitions. The push I feel in my arm, the hotness I sense in my finger, the blue I see, are not sufficiently "objective," i.e., are too strongly dependent on extraneous circumstances, on my own state of mind; in a word, they are unmeasurable in and by themselves. If a science were to be based on them that science would remain descriptive and primitive. Coerciveness and esthetic immediacy are inherent properties of these P-experiences, but systematic relations are not. Hence, to stabilize and render systematic the "pull" experience we set up an instrumental arrangement, perhaps a spring with an attached scale called a dynamometer, and allow the agent against which the force in my arm reacts to pull out the end of the spring. The mark on the scale opposite the end of the spring is the magnitude of the force, thus "objectified" and measured. In a certain sense, somewhat difficult to justify logically, this procedure *defines* the force, and the definition is called instrumental or operational. In the present terminology, it provides the passage from P to C.

The instrument performing this function for temperature

is a thermometer, for color it is a spectroscope (together with auxiliary apparatus), and so on. The use of instruments in the transition from P to C is very general, it is employed as a scientific procedure in the psychological and social as well as the physical sciences. The C-meaning of an individual's response to a stimulus, for instance the intensity of a heard sound, is not a subjective reaction (P-experience) but a location on some scale, constructed, perhaps, by an elaborate compounding of "just noticeable differences." Opinions (P-experiences) in a society are made objective (C) by the use of devices called questionnaires.

The scope and significance of instrumental definitions in science and elsewhere have been subjects of extensive discussion.[11] The claim that *all* scientific definitions must literally be of the instrumental type has now been largely abandoned; whether they are always operational in some wider sense clearly depends on how wide a sense is advocated. Detailed comments on this topic may be found in H. Margenau, *The Nature of Physical Reality* where it is shown that operational definitions form a special class of a larger set of epistemological relations called *rules of correspondence* or, in Professor Northrop's terminology, *epistemic correlations*. The latter are not purely logical relations, and certainly not identities, although they have often been mistaken for one or the other of these. They mediate between P and C experiences and

[11] See, for instance, "The Present State of Operationalism," a forum in *Scientific Monthly*, Vol. 79, No. 4, October, 1954. Also, H. Margenau, *The Nature of Physical Reality*, New York: McGraw-Hill, 1950.

represent epistemic elements of great importance. While far from being directly implied or dictated by the constituents of the P- or the C-field, they are still not arbitrary. Establishment of scientific theory involves a matching or mapping of P-experiences against a C-field by suitable correspondences. When a new P-fact is to be "explained," it is necessary to relate it to a construct *and* to specify the manner in which that construct is connected with protocol experience. E.g., to match the seen "blue" against the construct "frequency of an electromagnetic wave" is not enough, even if the equations controlling that construct (Maxwell's) are fully known. It is also necessary to supply operational definitions for electric and magnetic field strengths.

Instrumental definitions take us from a P-experience to certain simple concepts like force, temperature, and color. Science often requires further passages into the field of very abstract concepts, where meaning is related to P in remoter fashion. For instance, out of a number of quantities like the position and velocity of a body, all of which are operationally definable, science constructs the "state of motion" of the body and places it in a causal context with later states. Evidently, the epistemic correspondences connecting a state with immediate sensation are here more elaborate than operational definitions, but they serve the same purpose of linking a C-experience with P. By compounding simple rules of correspondence such as operational definitions with abstract relations, passages from P to very abstract and remote parts of the C-field can be established.

In the far reaches of the C-field, away from the P-plane, trans-empirical principles play an important role. For one asks of a theory more than that it shall merely provide an epistemic link with protocol experiences; while this is a minimum demand, the history of science shows it to be an insufficient criterion for acceptability, often incapable of deciding between rival complexes of constructs which are equally successful in explaining known empirical data. The additional requirements imposed are spoken of somewhat vaguely as economy of thought, Occam's razor, simplicity, elegance of conception, logical fertility, extensibility and so forth. Attention is given to them in *The Nature of Physical Reality*, p. 75, where they are called metaphysical principles. Despite their extreme importance for an understanding of modern scientific method—they have come to be relied upon with almost unjustifiable confidence in recent physics—I shall not review them here. It should be emphasized, however, that *in embracing these principles today's science includes within its proper methodology certain premises which most philosophers would regard as value judgments.*

Postulates and the Structure of Scientific Theory

The various vaguely defined components of our cognitive experience whose role was surveyed in the preceding section permit passages back and forth between any part of the protocol domain and the field of concepts. A few such "passages" will now be illustrated in some detail. Crucial for their correct understanding is the meaning of the following metaphors: movement in the cognitive field, passage from one component (e.g., P) to another (C). In fact this meaning differs when the problem of knowledge is seen from different vantage points.

Let us consider the relatively simple "passage" from subjective (P) force, as defined, to the construct force and thence to Newton's law, $F = ma$, which is a relation between that construct and two others. It can be studied from *four* different points of view.

1. *Psychologically,* to a person who knows physics, the "passage" is nothing more than a shift of conscious attention from one of these items to another. The connection is incidental; it is learned and then becomes a habit which controls thinking. It is furthermore reversible, and the order of the items is unimportant.

2. *Logically,* the situation is different. When the connection suggested by physical theory is fully established and accepted, that is, when one is willing to employ without question the operational definition of force and to assume the validity of Newton's law then, to be sure, the relation between the three components is one of complete entailment, and their order is again unimportant. But before physical theory has come to dominate the scene, each of the elements,

P-force, operationally defined C-force and the concept *ma*, has a private meaning of its own and these meanings are *not* logically related at all. The finding of operational definitions, the establishment of the law $F = ma$ are not logical procedures, for they depend on trans-empirical principles as well as on empirical verification. At the primary stage of cognitive experience, therefore, logic has very little to say concerning the passages in question except, of course, that it guides every step in the tentative procedures which lead to the establishment of theories.

3. From the point of view of a given individual's development, i.e., *ontogenetically,* little of general interest can be said about the movement under study. All normal persons presumably have an awareness of P-force although, unless they know something about its scientific significance—that is, unless they have an inkling of its connection with the other two concepts—it will remain obscure and commingled with other impressions. Then, as one learns physics, one sees the relation between the subjective force and a force operationally specified, and the former takes on a significance beyond the mere sensations of pushes and pulls. There is nothing unique about this temporal order, however, for it is perfectly conceivable that one might learn to recognize forces as indicated by spring balances even if one had no faculty for feeling pushes and pulls. Furthermore, some people learn first the operational definition of force and then the so-called law, $F = ma$; others learn these concepts in the reverse order.

Ontogenetically, therefore, the passages appear quite different than they do from the psychological and the logical standpoints.

4. Finally, one might view them in their *historical* occurrence. Which of the three components was recognized first in the history of science, and how were the connections between them discovered? In the present simple example the sequence goes clearly from P to C to *ma,* with the intervening connections emerging in that order. It should be mentioned, however, that in earlier theories, notably Aristotle's, the trend could be said to go from C to P. Very often in the history even of modern science the passage is from construct to observation: An elegant and attractive purely mathematical construct pleads for application, and sooner or later a scientist finds correspondences of the pure concept to P-experience. Many instances of such discoveries can be cited, the last, perhaps, being the correspondence between the vectors in complex Hilbert space and the states of quantum mechanical systems.

In the following context the term passage will not be used exclusively in the psychological, logical, ontogenetic or the historical sense. Its meaning shall be this. Suppose we *had* all P-facts within a given domain of experience. We may then ask: How do we select the correspondences, operational and otherwise, that will take us to the C-field? Having found the rules and established the C's, we ask further, how do these C's hang together among themselves? Here we rely strongly on logical and mathematical relations. The metaphysical principles mentioned in the previous section now become dominant, leading science to seek fewer and fewer constructs of explanation and to arrange them in a logical hierarchy. In a sense, then, P-experiences are primary; the movement proceeds—by definition—from there to simple constructs like C-force, thence to more complicated and more abstract constructs like *ma* and beyond these to very recondite constructs we have not yet mentioned in this connection. Just where it ends is a question we shall consider carefully below. If a word is wanted we might call this somewhat arbitrary order among the components of cognitive experience, which moves from the simple and unanalyzed to the complex and logically articulate, the *epistemic* sequence. As we shall see, it can also be reversed.

Let us now show, by reference to three examples, how science performs this passage.[12]

1. Nuclear physics is based on an increasingly large number of P-observations, associated with such terms as cloud and bubble chamber tracks, certain visual evidence of particle emission, scattering of nucleons, decay of nuclei and many more. In the strictest sense all these terms name concepts, but concepts so close to P that the reference to immediate observation is clear and present to every physicist. They are not

[12] The reader who is not interested in scientific detail may omit the three examples which follow without detriment to our general argument concerning the role of postulates.

historically primary, for many of these observations were suggested by unverified constructs which imaginative physicists had already conceived; nevertheless they are protocol facts, primary in the epistemic sense, because every theory of nuclear physics must be tested against them.

From these P-experiences science proceeds, often via instrumental definitions, to certain measurable quantities not directly perceivable but closely related to observation. Here it features decay constants, binding energies, atomic numbers, mass numbers, magnetic moments, scattering cross sections, reaction rates, spins, parity, strangeness, and the like. We are now definitely within the C-field; not far enough, however, to obtain the coherence (metaphysical principles) which we desire. Hence the epistemic process goes further, so to speak into the abstract regions of the C-field where it engenders, always subject to verification in a manner to be studied presently, increasingly imperspicuous concepts like elementary particle, a universal charge e, nuclear forces, spin operators, Hamiltonians, symmetries of state functions, tensor forces, scattering matrices, and a profusion of nuclear models, all endowed with logical relations and so chosen that the concepts of lower order are entailed by them. And since the latter are connected with P-experiences by definite rules, the abstract constructs just enumerated "imply" the observations. The word imply is used here, of course, in a rather special way.

The question now arises: how far can this systematic unifying process, this excursion into the domain of the abstract, be carried before our inquiry stops? No general answer can be given, for this part of the range is open. At any stage of scientific development there exist terminal constructs with relations sufficiently fruitful to form a theory, and that theory is then called a set of axioms or postulates. Science can never give assurance that a given set of postulates is final or that, since it may fail to be final, it is correct. Usually as science progresses, changes in the P-plane of experience, new factual discoveries, enforce changes in postulates to accommodate such novelty, and simultaneously there is a trend, made possible by improved theoretical understanding, toward merging several postulate sets into one, frequently by exchanging pictorability in mechanical images for that abstractness of conception which insures a greater logical range. One might say that science has two moving frontiers, one at the P-plane where gaps are filled and experience is broadened, the other far in the C-field at the place of postulates, where penetration of new logical and mathematical potentialities takes place for the sake of conceptual unification. But to return to our example: the postulates of nuclear physics, unsystematized at the moment because our knowledge is highly incomplete, and for the same reason somewhat contradictory, nonetheless contain such tentative commitments as: choice of Hamiltonians, charge invariance, symmetry of state functions, nonconservation of parity under certain conditions, and specific nuclear models. Here we find the second frontier, as yet unconsolidated; the existing ambiguity among nuclear models is a clear indication of the non-fixity of the postulates of nuclear physics.

2. The next example will also be taken from physical science and will be treated more briefly. Chemists and physicists have been aware of certain protocol facts related to heated bodies for about two hundred years. They know that when two bodies at different temperatures are brought in contact the cooler one becomes hotter, that the rise in temperature is determined by the masses ("heat capacities" in modern terminology) of the bodies, that gases often cool when they expand. Such observations have brought about many instrumental definitions: temperature as already outlined, quantity of heat, pressure, volume, entropy, enthalpy, free energy, etc. Again, we have now entered the C-field. But we go on in an effort to unify, introducing among these constructs certain theoretical relations which are expressed in the four laws of thermodynamics. These form the postulates of thermodynamics.

As an aside I might note that some textbooks call the laws of thermodynamics "empirical generalizations of a vast array of observations." While this statement emphasizes the general validity of the "laws" in P-plane experience, it is of course erroneous as a logical or an epistemic sentence. For it seems to say that the consequences of the "laws," which are non-denumerable and infinite because the "laws" are universal propositions, have been confirmed by the finite set of observations which have been made thus far. This, strictly speaking, is impossible. Taken literally, therefore, the textbook statement contradicts the elementary insights concerning universal statements of the earliest Western philosophers, notably Plato, and bespeaks the gulf which still exists between the languages of scientists and philosophers.

Thermodynamics reaches its postulates rather "early"; they stand fairly "close" to P-observations. By this we mean that relatively few constructs intervene between observations and the primitive, axiomatic propositions of the theory, in contrast to nuclear physics. There, the sequence from cloud chamber tracks to Hamiltonian operators is long and moves far into the abstract C-domain. In thermodynamics one defines such constructs as temperature, heat content, thence entropy and finds them at once related postulationally by the second law of thermodynamics. This circumstance has worried scientists over the last 150 years, causing them to wonder whether further abstraction may not allow the postulates to be unified or fused with others already employed in other disciplines. The last conjecture, fusion with others, has indeed proved feasible and is now in vogue. A subject called kinetic or molecular theory of matter, starting from its own set of P-experiences, had acquired, quite in accordance with the principles here outlined, its own set of postulates, and these were largely identical with the postulates of ordinary dynamics. If one adds one further, non-dynamical postulate to that set, namely Poincaré's ergodic hypothesis, perhaps in a more modern form (or, in a different guise, Gibbs' postulate of uniform phase density), the laws of thermodynamics, which were previously postulates themselves, become derivable, and hence merge with the kinetic set plus the ergodic hypothesis. In this instance, then, the postulational frontier is pushed forward by merger rather than by the creation of new ideas.

3. Social science employs the same epistemological formalism as the physical sciences, albeit perhaps in embryonic fashion. While this is not always admitted, it seems to me that careful analysis does bear out the parallelism we have traced. The problem will be studied further in Chapter II. In that chapter we shall analyze among others one famous theory of social behavior, Durkheim's theory of suicides, in terms of the present epistemology. Hence we forego the discussion of specific examples from sociology at this point but comment briefly on the axioms or postulates of that science.

Unanimity with respect to verification in the social sciences is largely lacking, and therefore the acceptance of any proposed protocol-construct-postulate sequence is far from general; but the underlying structure of all such proposals is evident. P-experiences, everybody will admit, are observations concerning the overt behavior of people. By certain epistemic rules these observations are consolidated into constructs like happiness, honesty, health, wealth, social integration and these are embodied in certain relations to be called *de facto* laws. (At the moment we are not raising the question whether law expresses what man *ought* to do, but profess to be legal realists; the normative question is fully discussed in Chapter III.) Some would regard these laws at once as postulational, thus making the epistemic sequence very short. Most of us probably desire to go further, hoping to derive the regularities codified in law from more fundamental axioms. In our society the last postulational recourse is usually taken to the avowal of personal liberty, respect of neighbor, and the like. Reflection will show, I believe, that every other theory explaining human behavior has a similar structure.

The word axiom (Greek: ἀξίωμα) means basic truth.[13] It is here used synonymously with postulate, a fundamental hypothesis or posit whose truth is *assumed*. There is no way of demonstrating the truth of axioms directly because they stand at the end of the epistemic range and are therefore unrelated to anything that might be logically antecedent. On the other hand, they are not arbitrary, since they can be verified through their ultimate consequences in P-experience, as we shall see. Since they have no ties to matters that are logically prior and depend for their validity mainly upon the details of P-experience, they are subject to being altered or even rejected in the flux of novel P-experiences. This feature makes science flexible, progressive, and self-corrective; science could have none of these qualities if axioms were bonded

[13] Euclid, in whose writings most readers have probably first encountered the term axiom, did not use this word at all. He employed the synonym *aitema*.

twice, once to absolute and prior truth and once to protocol experience. Indeed, in that case science would be impossible.

Induction and Deduction

Though the topic is slightly off the main route of the present inquiry, I shall now speak briefly of the psychology of scientific discovery and particularly of the problem of *induction*. This word has a double meaning. In the first instance induction links P-facts or, more precisely, concepts near P, with other such concepts in a manner avoiding any significant reference to the C-field. Having seen many black birds, all of which are crows, one might be tempted to say, all black birds are crows. This is a crude example of induction of the first kind. A slightly more scientific one is the inference that all men are mortal because every person in our experience has died, or the negative inference that there is no life on other stars because there can be none on our sun. In each of these examples a great but finite number of observations is generalized in the expectation, but never with the assurance, that the generalization is true.

Induction of the first kind is rarely pure, for there are usually rudimentary theoretical constructs which corroborate it. Thus when asserting that all men are mortal, one intends to say more than the truism that all men we know have died: one wishes to imply something about the process of aging which makes it necessary for man to die. This necessity comes from an appeal to theory, to the constructional equipment already at hand. Similarly, when people base their disbelief in miracles on simple induction—saying that miracles cannot happen because they have never seen one—they are in fact often asserting that their positive knowledge about laws of nature precludes miracles.

More interesting in the present context is the second meaning of the word induction, which designates a form of passage from P to C. A given set of observations, to be accommodated by reason, must be supplied with epistemic correspondences to concepts; these must be organically related by laws and principles, and the whole connective tissue in the C-field extending to the axioms must be filled in. Ordinarily this process, which as we have seen is not guided by clear logical rules, is performed gropingly through trial and error, details being inserted step by step on the way. Only occasionally, by faculties rare among scientists but akin to competences which Professor Pitirim Sorokin calls supraconscious, a blessed individual performs the passage in a single creative act. Starting with unorganized data, he will suddenly see before him an elegant explanation with all its attendant linkages to the facts and be convinced of its adequacy. This successful vision, called the inductive leap, does not exempt him or others from verifying the theory through a subsequent, slow, and sometimes painful process. Indeed such inspirations have often been shown to be wrong. The inductive leap in science is the counterpart of inspiration in poetry, revelation in religion; the difference seems to be

THE METHOD OF SCIENCE

only that science is able to prove its inspirations right or wrong by independent means.

Induction of the first kind might be called descriptive, the other explanatory.

The question is sometimes raised whether science explains or merely describes phenomena. If explanation is sought in terms of metaphysical ultimates, science does not explain, nor does any other human discipline. But science certainly goes far beyond description. Hence, to give proper meaning to the word explanation, we should associate it with progressive excursion into the C-field and with the increase in logical power which attends this excursion.

To explain the motion of a falling stone I appeal to Galileo's law: the stone falls because it suffers a downward acceleration of 32.2 ft./sec. This is a rather unsatisfactory explanation amounting to little more than description, but it does allow the phenomenon in question to be seen as an instance of a law of greater logical power. I may go on to ask, why does the stone have a downward acceleration? This second "why" is satisfied by reference to Newton's law of universal gravitation. A second explanatory step has thus been taken which includes the minor law in a major one. Next, the "why" respecting Newton's law receives an answer in Einstein's theory of general relativity; but here we stop, for that theory is today's postulate. We confront here a *chain* of explanations, and this situation is typical throughout science.

If, then, we take description to mean a verbal or even a quantitative account of P-experiences, retaining the word explanation for the progressive subsumption of minor facts, laws, or theories under major ones as just illustrated, science does both, describe and explain.

Induction, the passage from P to C, is an uncertain process; the data do not strictly imply the concepts. *Deduction,* which is the opposite procedure, is strict and certain, but it can only be performed when the C-field is full. Its results need not be correct, however; i.e., it need not encounter the protocol experiences specified by the conclusion.

There are some textbooks which still claim that science results from the inductive approach to knowledge. From what has been said the error of this statement should be evident; science uses both induction and deduction, the former process mostly in forming new theory, the latter in making predictions from established theory.

Verification

The difference between a conjecture or hypothesis and a theory is this. A theory is a conjecture which has been successfully confirmed and therefore "verified." Confirmation and verification will here be taken as synonyms, although the literal sense of confirmation is more nearly in keeping with the spirit of science than is verification: no matter how often a theory is confirmed, it can never be said to be at any time absolutely true; later refined observations will almost certainly alter it. Nevertheless scientists call a theory true so long as it has not met with any disconfirmation or refutation. What, then, is involved in the verification of theories?

Roughly speaking, to verify is to record agreement between the consequences of a theory, which as we have seen is itself an explication of more basic postulates on the one hand, and P-experiences on the other. Two questions thus arise. First, precisely what are the P-experiences which are eligible for comparison? And second, what is meant by agreement? Neither is given the attention it deserves in most treatises on the philsopy of science and yet, as we shall see below, each contains difficult problems.

In subsequent chapters I shall try to show that parallel problems regarding the nature of confirmatory evidence and the meaning of confirmation arise in ethics.

Turning to the first question, one is at once struck by the vast range over which P-experiences may extend. We defined it as that component of cognitive experience which is coercive, indubitable, or simply given. Nearly everybody acquainted with scientific method identifies what these phrases convey with sensory data, external observations, and the results of experiment. Now these are certainly included in, but perhaps do not exhaust all P-facts. Who doubts, for instance, when he is suddenly possessed by a feeling of joy, or fear, that these moods are coercive, indubitable, and given? These sentiments might therefore in a true sense be regarded as "data." The same applies to a sudden rational insight, a flash of mathematical discovery, the sudden remembering of an event by association. Yet none of these is ordinarily admitted to the protocol class of scientific experiences, at least in natural science. Interestingly, one of the claims of Husserl and his disciples, the phenomenologists, is that natural science allows itself too narrow a range of confirmation, that it should take cognizance of the vast domain of "eidetic introspections" which is open to reflection and suggests truths as valid as those conveyed by our senses—but science has so far denied this claim. Its denial is based primarily on the subjectivity of introspective moods. More specifically, many scientists insist that introspective data must not be used as criteria for the correctness of theories *because they cannot be shared.*

On closer analysis, however, it appears doubtful if sharability, intersubjectivity of a specific datal perception is a necessary attribute of protocol experience even for natural science, since unique observations are not ruled out as confirming evidence by the code of science. The records of individual astronomers are often used to check calculations, results of a difficult observation tend to be accepted without repetition; if a single scientist saw the abominable snowman of the Himalayas, his testimony would surely create an objective stir. But the feeling of terror that might strike him on the same occasion does not qualify in the same way as a scientific protocol experience.

Examination shows,[14] I think, that science admits as evi-

[14] H. Margenau, in *The Nature of Physical Knowledge,* ed. L. W. Friedrich, S.J., Marquette University Press, 1960.

dence single protocol experiences of the perceptual type if they are hard to come by or if, in the nature of things, they are not repeatable. Nevertheless, it is on the whole a communal affair; this follows, not from a special axiom of intersubjectivity but from its drive to explain as wide a realm of nature as possible. If this realm is taken to include other persons with reactions similar to my own as well as the class of so-called physical phenomena, consistency requires a maximum of sharability. For clearly, to assume that others have the same experience as I on the same occasion, a thesis scientifically confirmable in numerous ways, at once enlarges the available P-domain and provides opportunities, not restrictions, of which science readily avails itself.

Our first question was: which P-experiences are eligible for comparison with theory? We have so far answered it by saying: mine and others, provided they are of the perceptory ("cognitive") type. It remains to inquire why emotional factors are largely excluded. The reason lies in the purpose of natural science, which is to understand, anticipate, and predict external events in space and time. One sees here no willful discrimination against certain kinds of human experience but judicious restraint in selecting precisely those factors which are most likely to assure the success of its particular cognitive endeavor. Insofar as psychology deals with feelings, it often extends its interest to introspection, and if there is ever to be a science of human emotions, of fear, love and passion, it is quite certain that the customary P-plane will have to be extended. On a lesser scale, the lack of fixity of protocol experience is already quite apparent within natural science itself: the material with which the chemist confirms his hypotheses is different in detail from that of the physicist and the botanist; the data of the nuclear physicist today do not resemble those sought fifty years ago. Comparison of the very appearance of a modern scientific laboratory, in the glory of all its electronic and automatic equipment, with the glass-laden workshops of an earlier generation demonstrates this point. It is as though the infinite domain of all possible immediate experience, which is potentially available as P-experience, were being scanned by searchlights under the control of ever-changing scientific interest and theory.

Hence, we conclude: the protocol material with which theoretical speculation seeks contact is intrinsically unlimited, but the purpose of science as a whole, its character of being a cognitive enterprise, and the narrower goals of specific sciences, enforce selections which, at any given time, rule out large portions of immediate experience as irrelevant.

Before turning to the second question, let us cast a quick glance ahead at the principal subject of this book, which is the method of ethics. That discipline is often defined in terms of its goal or goals, is said to be, for instance, an expression of the human striving for happiness. Now this is certainly correct. But it is woefully insufficient, as we intend to show. It is no more adequate as a characterization of ethics, than is the following statement as a definition of science: science is an expression of the human desire for predicting phenomena. A proper philosophy of science must augment and amplify this common dictum, and such amplification involves a study of what precedes prediction, i.e., the postulational and theoretical structure of science. We shall find analogs for all these elements in ethics.

The second question is: What is meant by agreement between theory and the kind of P-experience relevant in a particular science; what are the details of the verifying act called confirmation? At first glance the answer seems obvious: theory predicts something and observation finds it true. Unfortunately this uncritical view of confirmation is completely erroneous, for observation practically never supplies precisely what theory specifies. One looks for agreement only within a certain range of tolerance or error, and thereby hangs a tale for students of ethics, which will later be told.

Suppose we have a theory of celestial mechanics, and on the basis of previous measurements a star should rise above the horizon at a certain calculated instant of time. Careful observation shows it to appear six milliseconds later. Naively this represents a refutation of celestial mechanics, yet the astronomer will nevertheless regard it as a confirming instance, claiming that the deviation of six milliseconds lies within the tolerance or the probable error of his telescope-time piece combination. Appeal to errors here and in all scientific measurements reduces the severity of the verifying test or, in a more fundamental sense, makes the verifying test possible.

To see the meaning of what the scientist calls error, and to confute the popular belief that error is a human failing, let us first note its inevitability. There is no exact empirical "truth." What is the true weight of this book? You might put it on a scale and weigh it; the result will be a certain number of ounces, drachmas, grams or milligrams, but it will not be precise. Nor will it help to weigh the book on a microbalance—which will doubtless be damaged in the attempt. The most accurate scale useful for the purpose gives an indication within the "least count" of the instrument, which might be, say, ten milligrams. But upon realizing this it is still incorrect to suppose that the true weight is known within that range of accuracy, for if the measurement is repeated different answers result. Only a very crude device like a grocer's scale produces the illusion of constancy, all good measurements scatter. Indeed the scientist's method of finding the weight of the book, in case he were interested, is to weigh it on a good balance many times, to record all results and then to take their arithmetical mean. In this way he "averages out" the incidental errors and arrives at the "truest" value available from the actual set of measurements. And he would prefer to use the superlative rather than the positive "true," for he knows that if the series of weighings were continued a different and perhaps "truer" arithmetical mean might result. But he has no absolute assurance that the series of true values converges to a limit as the weighings continue, especially when more and more refined balances are used.

To define empirical truth it is necessary to establish a range of tolerance. That range is often called the probable error, with the misleading connotation of evitability. In fact, the blame man assumes when he speaks of error in this context is an inevitable feature of all P-experiences, an intrinsic vagueness or haziness of immediate sensation manifesting itself in this and other ways. Tolerance is by far the better word, but I shall follow custom and call it error. Now there are mathematical theories which relate the probable error of a sufficiently numerous set of measurements to their observational dispersion, to what is sometimes termed their standard deviation. Such theories rely on further postulates, primarily one which assumes the results to have a "normal" or "Gaussian" distribution, but these details need not concern us here. What matters is this. The definition of probable error and its use as the limit of tolerance in observations—and in establishing empirical truth—are deliberate, arbitrary, and conventional. The method employed in evaluating a set of measurements and also in assessing the reliability of one measurement, stands and falls upon acceptance or rejection of principles not provided by the measuring process itself, nor by the theory to be tested.

To see this more clearly, let us return to the astronomer's observation of the rise of a star. If the interval of six milliseconds were deemed to be larger than the probable error of his apparatus, he would have to reject his theory, if smaller he would retain it. But the choice of probable error is largely arbitrary, as we have noted. It follows that, if the tolerance were set at zero, no theory could ever be verified at all; if it were taken as ten times the probable error, many theories now considered false could claim validity. The very possibility of practical confirmation depends on commitments extraneous to the measuring process and to the theories under examination.

Lest I overemphasize the arbitrary nature of the statistical postulates which enter here, let it be said that they are far from capricious or dogmatic. It is their logical status which concerns us here, not the manner of their imposition. There is universal agreement as to these statistical postulates, agreement springing partly from the very success of the scientific method that employs them, partly from the simplicity and elegance of their mathematical formulation. The remarkable development of the modern theory of probability illuminates with increasing clarity two facets of the verifying process: 1) the special nature of the statistical assumptions which this process requires, and 2) the grounding of these assumptions in reasonable axioms of their own.

Clearly, then, to verify a theory presupposes matters not affirmed by theory nor by observations. One must *select* a certain kind of P-experience as suited for the process of verification, and one must invoke *principles,* like the proposition: the observational tolerance equals the probable error of the measurement or the apparatus, to render possible such agreement between theoretical preduction and observation as science demands.

Thus we finally see that certain kinds of commitment, logically anterior to the practice of science, enter prominently at *two* places in its methodology. The usual postulates of scientific theory dominate its starting point; other posits, equally postulational and concerning what is meant by verification, are needed at the end, where theory makes contact with P-experience. In the sequel we shall continue to call the former the *postulates of science,* to the latter we shall for convenience give the name *principles of verification.*

The principles of verification were not born with scientific method. At first, validating procedures were crude and required none of this sophistication. Measurement was a look-and-see-affair and agreement was obvious. As precision grew both in the theoretical and in the observational field, the need for principles of verification became increasingly manifest, and in our day they have become matters of intense interest to the specialist. Postulates and principles of verification in science, I hope to show, have counterparts in ethics. They will be called *codes* (or *commandments*) and *principles of validation.* There, too, the latter have gone unrecognized until a certain stage of development of ethical methodology was reached. The point we shall make in due course is that ethical theory, like science, is a defective undertaking unless its involvement with these two postulational ingredients is clearly recognized.

II

The Method of Physical Science and Human Affairs

ABSTRACT

Six common arguments purporting to prove that scientific method is inapplicable to human affairs are presented and analyzed. Their import seems to be at variance with the actual success which application of scientific method to sociology, in particular, has attained in several instances. Hence we reappraise the initial arguments and find them largely wanting in rigor of conception and in cogency. Recent physical science has had to cope with difficulties very similar to those in the social sciences and it conquered them by abandoning its old mechanistic, single-event predictions, contenting itself with the forecasting of averages in ensembles. A similar style of analysis and prediction is to be expected—and is indeed present—in the social sciences. Ethics, too, must place its accent on collective behavior.

The Problem

To say that all knowledge, let alone all experience, owes whatever reliability it seems to possess to the epistemological processes outlined in chapter I would be a vast misrepresentation, for it is clear even without analysis that most items of experience, and especially the intricate and vital ones, defy the pattern of classification which recognizes only data, concepts and rules of correspondence; science cannot deal with them. An honest appraisal of the grounds on which men in the present scientific age base their actions, their decisions, the routine of their lives, concedes science but an infinitesimal area of direct control: the large amorphous matrix of human experience contains only a small region which is occupied by the crystalline structure we call science. To be sure this region grows in size as time goes on, but there is no evidence to be drawn, either from the nature of the scientific process or from the historical development of science, which faintly suggests the likelihood that some day all human experience will be organized in the manner we have described. On the contrary, it seems as if the crystalline, scientific portion of knowledge, in terms of our metaphor, while growing forever, will always be small in comparison with the potentially infinite domain of man's total knowledge. Science is precious, not because of any claim it makes to universality, but because its successful function in human experience is rare, uncommon, in some respects miraculous.

Where, then, does the method outlined succeed with greatest frequency? Evidently, in the physical sciences, where it originally evolved and has now gained its finest articulation. Precisely this fact, however, raises suspicion with respect to innate limitations of the method, poses the question whether an artifact designed to organize our understanding of physical phenomena, that is to say, lifeless unconscious processes, has any chance of succeeding in the sphere of living, conscious beings; whether a scientific account of the transactions of men, of their social and ethical behavior is possible at all. If not, our attempt to discover a scientific structure in ethics must fail.

There is indeed a large body of opinion which maintains this pessimism, maintains it with unconcealed delight over the spiritual incompetence of science. In the present chapter we review the reasons commonly given for this supposed predicament. Wherever possible, the weaknesses of the arguments marshalled in support of the spiritual bankruptcy of science will be exposed, yet on the whole we shall remain short of proving that science is *necessarily* applicable to human affairs. This is an impossible task, fraught with fundamental difficulties which it is well to summarize at the very start.

The question is what it means to demonstrate that a program, hitherto unfinished, can or cannot be carried through with success. It appears in a variety of settings. In its simplest terms the question arises in elementary logical or mathematical situations, where the language and the conditions are completely fixed and no new knowledge can affect the problem. Thus, for example, machines can decide whether a certain logical proposition is right or wrong in the sense of ensuing or not ensuing from stated premises. More interesting is the case where the problem is posed but the answer is not known, and where one desires to know whether the problem has, or does not have, a solution. This occurs in the logic of decidability theory and in a wide area of mathematics. A given differential equation may be extremely difficult to solve; nevertheless it may be possible for the mathematician to prove the existence or nonexistence of a solution in terms of analytic functions, even though the solution itself has not been obtained.

A kind of existence proof establishing the possibility of some kind of scientific ethics, unknown in its details, is what we here desire—but cannot have. The reason is this: Logicians and mathematicians deal with essentially static situations, with conditions in which the language, the postulates and the rules of operation are completely fixed. In these areas new facts and additional postulates do not emerge as the process of solution goes on; the search proceeds in a stationary universe, it knows that elements already inspected will retain their character intact, it can classify, eliminate, arrange, and thus succeed.

As we pass from logic or mathematics to empirical science, which is exposed to the living flux of the world, the state of affairs is radically altered precisely because of the continued emergence of new factors of evidence that cannot be foreseen. Many scientists today are seriously engaged in efforts to create a "death ray"; to be more specific, they wish to design a system which will destroy or damage approaching missiles at considerable distances from their targets. Staunch traditionalists sneer at these efforts; having been brought up in a school which took the laws of nature as completely codified and known, they apply the rigor of mathematical reasoning and conclude that the search is fruitless, that the death ray can be demonstrated to be unfeasible. Yet the quest continues, carried on mainly by younger men who live amid the rapid change in physical knowledge and know that a refutation based on the principles of ten years ago is not a refutation today, nor one of tomorrow. And if history teaches a lesson, the young men are right: there is every justification for disregarding old arguments among living issues, especially when the end is strongly desired.

Yet there is a way of legitimately curbing the search for a death ray. If it can be shown that its creation contradicts some very fundamental principle, like the law of conservation of energy, then the effect will subside among scientists, and the issue will be regarded as closed. Strictly speaking, even this does not constitute a total annulment of feasibility, for it is conceivable that the principles of conservation may some day be found in error; but scientists with historical perspective know that a principle as basic as the one here in question is not *likely* to be abrogated at any foreseeable time. Pointing to a contradiction of some basic principle, therefore, constitutes *effective,* if not ultimate refutation. There exists, then, a way of practically *refuting* the feasibility of a scientific program.

However there is no way of *proving* feasibility short of success itself. The motivating power of positive research is not its certainty of success; in large measure it is the challenge arising from the possibility that it may fail. Because results are not assured, their worth when attained is greater. The other factor sustaining research efforts in the absence of guaranteed success is the strength with which results are desired. The case of the death ray here cited illustrates these things in simple, timely, but brutal fashion. Many other current efforts might have done as well and perhaps with less offense to the philosopher; among them one might mention the production of large-scale nuclear fusion in the laboratory, the construction of rocket motors for space travel, the creation of planetary atmospheres capable of sustaining human life, and, to go to the heart of the matter, I hope that the establishment of a universal code of ethics may be included among them.

But here we are soaring from projects that are strictly scientific to one which is philosophic and transcends science. What this means is that in the attempt to prove feasibility, we are certainly no better off and must expect to encounter at least all the difficulties which attend the task in empirical science. Existence proofs, logical decidability procedures are not available, and the only clear-cut verdict can be established by conducting an effective *refutation* by appeal to fundamental principles, as in the case of natural science. The only *positive* proof is in doing, i.e., in establishing a scientific code of ethics which actually works—a goal clearly beyond the aims of this book. Hence we must fall back on refutation, which produces evidence of a weaker form.

We thus find ourselves in a position that can be described as follows. We wonder whether the establishment of ethics in conformity with scientific methodology is a feasible task, or, more generally, whether the scientific method can be applied to human affairs. This larger context contains our more specific problem, and since the larger question has received an even greater amount of attention than the problem of scientific ethics, we shall also confront it here. Drawing a lesson from the preceding considerations, we forego every attempt to establish a positive answer. We survey what refutations have been offered in the past. Thus we discover several important arguments which claim finality largely by affirming that human affairs in their complexity, their intrinsic unpredictability contradict the most basic postulates of the method of science, in particular the method of physical science. These we shall try to disarm by showing them to be inconclusive, notwithstanding their popular appeal. If the attempt succeeds the problem is reopened, the challenge restored. Positive proof of our thesis beyond this is impossible.

But a measure of strength is given to our position if the following further point can be made. In their groping for satisfactory control of human interactions, for universal moral standards, for acceptable international laws, men have in fact, perhaps without being fully aware of it, employed the resources of the scientific method. They have already imposed the kind of validation upon these forms of control, upon these standards and laws, which scientists employ when they verify their theoretical predictions. In later chapters such will be shown to be the case in detail. The next section presents the customary refutation arguments in their strongest forms. The following section will then review some of the actual results which an application of science to human affairs has thus far achieved and thereby raise some incidental doubts concerning the cogency of the arguments to be put forth. Following that review we shall show that each of the arguments, when carefully examined, fails to prove its point.

Six Arguments Purporting To Prove That Scientific Method Is Inapplicable To Human Affairs

1. *The inherently destructive effect of scientific analysis.* When a chemist wishes to determine the chemical nature of some material, he subjects a sample of it to tests that ordinarily destroy it. This is the price he pays for accurate scientific knowledge. The biologist often kills an organism as he identifies it or studies its function. Science was once indicted, and alarm rose, when men spoke of vivisection. At present, archeology is at the threshold of becoming a science by virtue of a new technique called carbon dating, but there is some apprehension on the part of archeologists because their precious finds must be subjected to mass-spectrographic analysis.

It was long thought that physics was exempt from the curse of having to destroy in order to gain accurate knowledge: bodies can be weighed, motion can be studied, formulas can be written without detriment to any object. Indeed it seemed that in this area activities were happily constructive or creative; wonderful devices could be built out of crude materials for the benefit of man—yet it was supposed that only inanimate materials could thus be ennobled by the processes of physics. This exceptional state of affairs in physics must now be questioned, since it has been shown that on the most fundamental level, where the elementary constituents of physical nature reside, the destructive effect of scientific inquiry reappears. The customary interpretation of Heisenberg's uncertainty principle connotes that every act of measurement, indeed every observation of an atomic quantity, inasmuch as it must be performed with finite quanta of energy, produces a finite effect upon the system under observation and thereby *alters* its state in unknown and unpredictable fashion. The process of attaining scientific knowledge destroys that knowledge in the act of acquisition. While chemistry and biology alter, and usually degrade, the objects they study, physics leaves the objects intact but interferes with and often falsifies what they are doing through its process of inquiry. Physics might be said to be epistemologically destructive.

All these facts are sometimes summarized in the form of a massive accusation of science, phrased most eloquently by modern existentialists who claim that science poisons being in the act of apprehending it. As one reads Heidegger one senses that he looks upon "Sein" as something peculiarly indigenous to the inarticulate reaches of human experience, something conveyed to us at once by virtue of its primary quality, existence, and as such never in need of analysis. Scientific analysis, when performed, can yield distorted glimpses of specific aspects of being, but it is to be likened to the chase of a delicate animal which one should prefer to observe in its native habitat without molestation and which, run down and caught, no longer presents the true characteristics of its living situation. Science pursues its quarry relentlessly, and when it captures it and thinks it has the living truth, it holds more often only the corpse of truth. Many scholars of a humanistic bent share this view, and from it springs the uneasy feeling many of us have about the prospects of a world dominated by science. Its consequences for the problems of ethics are clear: you cannot analyze human actions in a scientific way without violating their true ethical essence.

2. *Science deals only with the quantative aspects of experience and leaves out qualities and values.* Human relations, actions, moral judgments are not quantitative matters; they cannot be caught up in numbers and equations: They are essentially qualitative aspects of experience. Science, on the other hand, is bound to rely on measurement and simple observation for its primary data, which are therefore *numbers*. Any doubt about this is at once dispelled when the concept of measurement is analyzed; measurement is a comparison of an unknown magnitude with a known standard, e.g., the ascertainment of how many times a yardstick must be laid off to cover a distance, how many ticks a pendulum clock makes in a certain time, how many marks on a protractor are contained in an angle, how many milestones a moving car passes in an hour. All these are numbers, because the ratio of one magnitude to another standard magnitude is always a number. Numbers go into equations, and equations are quantitative relations which sacrifice the qualitative aspects of experience.

To make quantitative treatment possible, science tends to *convert* qualities into mathematically tractable quantities. It cannot content itself with the sensed, intuitive blue of the sky but is forced to convert it into a wave length, the sour taste of vinegar must be quantified into a pH number, a harmonious chord into frequencies, a noise into decibels. Our first argument holds this conversion to be destructive, degrading, repulsive and therefore objectionable, the present one declares it insufficient as a description of reality, maintains that it leaves certain important matters out of account and thus results in a record that is incomplete.

Historically, this issue arose spectacularly in the controversy between Goethe and the Newtonians in connection with the problem of color. One can hardly resist the appeal of Goethe's eloquence even if one is willing to grant the poet's ineptitude in experimental matters—for there seems to be some evidence that his failure to reproduce Newton's spectrum of white light was occasioned by the absence of a slit in his spectrometer. The visitor to the Goethe museum in Weimar, who sees the exhibit of color samples and reads Goethe's rich and impressive verbal description of the psychological effects of these colors is easily convinced of the merits of his point of view. To fix it in terms of a specific example, I recall here the claim of an artist who lectured on the "essence of color" some thirty years ago. The fact that color is a quality, he said, can be demonstrated as follows. He explained that the physicist describes visual color in terms of two quantities, hue (wavelength) and intensity (energy reflected per cm^2 per sec.). Then, on the same canvas, he exhibited two contiguous areas of painted blue which were noticeably different to the eyes of all observers. Nevertheless he said, correctly, that the two colors were of the same wavelength and the same intensity according to the best measurements with physical instruments. Something evidently escaped the scientific grasp; the artist's point was made: Color in its qualitative integrity refuses to be caught in the act of scientific analysis. The transfer from so simple a situation to the behavior of men, which involves reactions to far more subtle experiences than color vision, seems fairly safe and causes the hope for a scientific treatment of ethical problems to recede, perhaps beyond recovery. We shall, however, return to this instance later in this chapter.

3. *The complexity of the living world.* When science, even in the physical world, comes to grips with complicated phenomena, its power appears to wane. At present, there is no satisfactory theory of turbulent motion; meteorology and oceanography are not capable of precise predictions and the chemist, confronted with a fluid possessing innumerable degrees of freedom, is at a loss to theorize convincingly about its melting point. As one ascends the scale of complexity and considers biological systems, the usual types of explanation become progressively more precarious and new methods come into play, methods calling for the abandonment of causal description in favor of holistic and teleological theories

which in the eyes of many philosophers, vitiate the scientific enterprise.

The meaning of complexity in these examples resides in the excessive *number of variables* required for a full description of the processes under study. To explain or predict the motion of a stone, a projectile (neglecting air friction) or a star, one needs, aside from the basic laws, nothing more than a knowledge of two variables: position and velocity at a given point. In the case of a physical field, a continuum of numerical values, again of two kinds, is required for a satisfactory account of its behavior. Contrast this with the simplest instances in the biological world, e.g., the explanation and prediction of the growth of a newborn organism. The problem at once becomes staggering, for not only is the number of variables to be determined very large, it is indefinite; no one can say what factors should be measured for the sake of predicting growth, nor is there a satisfactory law in terms of which such predictions could be made. The course of a disease even in the simplest cases is subject only to statistical prognosis, and less is known about normal mental development and mental aberrations. Concurrent with this transition from the physical to the biological sphere is a change in motives for research, a change from curiosity and the drive for understanding in the physical sciences, to therapeutic and humanitarian desires in the "art" of healing.

The difficulties are multiplied enormously when science

turns from individual to social matters, where the variables required for adequate description pass completely out of sight and defy enumeration. Relations of affection, hatred, trust, obligation, and many others, which have no meaning for isolated individuals, now enter the scene and these, even if they were capable of being quantified, will surely transcend the methods of itemization practiced in science.

4. *The difficulty of control.* Science is at its best when it is able to *control* its variables. One view, a mistaken one, I think, affirms that science develops *only* where control is possible. More credible is the thesis that laws take on a particularly *simple* form under conditions which may not always hold in nature but can be created or contrived, that these conditions are exemplars which allow an approximate understanding of actual happenings. The thermal expansion of gases may be cited here, to wit: If a gas in a container is heated with consequent changes in its pressure, volume and temperature, these three variables are related by the so-called equation of state of the gas. In general, that equation is a highly elaborate one, containing many constants and expressible only in an infinite series of terms. For some gases, the higher members of the series are small—these gases are called ideal or near-ideal—but in general, especially under conditions just before condensation occurs, all terms in the series are large and the law is unwieldy. At low pressures and large volumes, however, every gas obeys the ideal law, a very simple equation (pressure $\times$ volume $=$ constant $\times$

temperature) which is the pride and joy of the chemist. By controlling the volume, i.e., by expanding the gas so that a small number of molecules occupy any given volume, the miracle of simplicity is enacted.

The literal transcription of this state of affairs to social science, let alone to ethics, is ludicrous. If one were to say that sociology has complex laws because society is too populous, that its laws take on a simpler form if more and more people are removed from it, he would in effect be saying that society has no laws at all, because removal of persons alters the structure of the group to be studied. Or if the statement meant that sparse communities are in principle easier to understand than congested ones, I would doubt its correctness. At any rate, we are unable, in a rational society, to alter a community by dispersing its members, we lack the facilities for control even if they did make the laws of sociology simpler than they are in the actual setting of societies. Here, it is often felt, we have come upon one of the major failures of scientific method in its application to human affairs.

5. *Sentient beings can make decisions, scientific objects cannot.* The whole of science is based upon the causal postulate which alleges that the state of an entity, once given in its characteristic features, determines completely the evolution of the entity. Very few branches of science justify the full rigor of this postulate, yet its power in all conforming instances is most impressive and suffices to convince the ex-

perts of the universal scope of the causal principle. Were it not for the complexity of phenomena and for the difficulty of controlling some of them, they hold, causality could be recognized everywhere. Let us, for the sake of argument, accept this reasoning. One can then see quite easily that it must be confined to inanimate objects, or to inanimate and conscious objects devoid of free will. Any being capable of making a free decision destroys the causal evolution of states by that very act, since freedom contravenes determination. Men are free agents. To believe that science is applicable to human actions, above all to ethics, then seems to imply a denial of freedom.

In a sense this conclusion opens itself to empirical test. If it is true, the behavior of citizens under tyrannical rule should be more amenable to scientific treatment than actions in a free society, predictions in an authoritarian society should be easier and more exact. There is, I think, a good deal of evidence for such a claim, and the force of the argument under review is greatly increased by it. But there is a correlated conviction frequently encountered among those who are susceptible to the suggestion that science, to be applicable, presumes absence of freedom. This is the belief that science *engenders* constraint, impairs or nullifies freedom, and it is totally erroneous. No dictator was ever a scientist, and rarely did science put a dictator into power. There is an interesting exception; Northrop (*Meeting of East and West*) shows that the Diaz dictatorship in Mexico was the product of the Comtian theory of science. Note, however, that Comtian positivism is not science in our sense. On the contrary, the lesson of history clearly is that science and tyranny do not get get along together, that inquiry, including causal inquiry, must be throttled before a totalitarian regime can succeed. Let us return, then, to the lesser conclusion according to which a science of human behavior or human relations commits one who believes in its feasibility to a denial of free actions.

6. *The transaction between knowledge and fact.* Physical objects are indifferent to the knowledge man acquires of them; they display a kind of stability or inertness which makes predictions possible. Stars continue to move in a manner expressible by laws whether or not these laws are known, whether or not predictions are applied to their motion. The knower stands aloof from the known, the object is epistemologically isolated from the scientific process. What was said about stars clearly holds in essence for most occurrences tractable by physical science, indeed for all processes involving objects lacking consciousness and the ability to react to knowledge.

In psychology the state of affairs is different. If a subject knows what a psychologist knows about it, it may spontaneously act to make the latter's knowledge false. There is, furthermore, the possibility of simulation or of making erroneous answers to an inquiry, deliberately and otherwise. The manner in which a doctor or a psychiatrist asks about the state of a patient's health may induce a pain which is only a subjective reaction to the momentary situation, without relevance for the actual state. Here knowledge and fact become intertwined, objectivity ceases, and scientific analysis is bound to fail.

While on the level of individual behavior this transaction between knowledge and fact remains erratic, it takes on systematic forms in economics and the social sciences. Here we encounter the possibility of self-fulfilling and of self-defeating predictions. Knowledge that a bank will fail, if widely held, with cause or accelerate its failure; the knowledge is self-fulfilling. So is the prediction that the price of a stock will rise or fall, but if the prediction were made more quantitative, if it were claimed, for example, that a certain stock would fall by five points at a specified time, the rush of selling might bring it down below the predicted value and the knowledge would be quantitatively self-defeating. Advertising and propaganda efforts skillfully seize upon the opportunities inherent in self-fulfilling claims, and good statesmanship tries to avert danger by announcing self-defeating predictions. The psychology of nuclear deterrence in international relations treads a thin line, let up hope on the safe side, of the boundary between self-fulfillment and self-defeat.

It is not our purpose here to stress the importance, nor the danger, of social processes which effectively entangle knowledge and fact; we simply note their existence. The claim is made that they obliterate the demarcation between the knower and the known, remove the distinction between the world as spectacle and man as the beholder, engage man in an active role in the drama of events not only by virtue of his deeds but even through his attempts at knowing and predicting. On the face of it, this claim is good. It points to the epistemological counterpart of what in science is called feedback, a subject long known as a curiosity but developed, mainly by Shannon and Wiener, into a theme of vast importance in science and in philosophy as well. I deem it worth a little attention even in the present limited context.

The performance of an engine without governor is determined by the supply of fuel it receives from the outside. When equipped with a governor, it controls its own behavior because the governor regulates its fuel supply. This can be arranged in a variety of ways. The normal one is to throttle the supply as the speed of the engine grows, an extremely abnormal one is to increase the supply as the speed increases. The engine will run at more or less constant speed in the first instance; it will destroy itself in the second. In either case it "feeds back" some of its mechanical output into its own performance, stabilizing itself in one, rendering itself unstable in the other instance.

The analogy with the prediction problem is evident. In the physical sciences the universe is supposed to run without control by knowledge, using only resources provided by the creator. In the social world, knowledge is fed back into the systems studied, and there is stability—if you please —in self-defeating predictions, instability in others; although there may be some question here as to the propriety of the term stability. Notice, however, that feedback in its strict scientific meaning couples physical output and physical input, whereas our argument is concerned with the coupling between knowledge and facts. Hence, when the word feedback is used in this connection, it is at best a metaphor. Our question is whether this transaction defeats scientific treatment of the facts, and this question is answered affirmatively by the skeptics. Certainly, an affirmative answer on the stage of social science annuls all chances of success for science in the moral world, where the interrelation between knowledge, ethical norms, environment and actions are infinitely more subtle than in descriptive behavioral science.

Physical Laws and Social Behavior

If the logic of the six preceding contentions is sound, physical models and the type of laws encountered in physical science cannot succeed in sociology. Yet there are known examples, sporadic to be sure and not generally accepted, which promise possible success. I do not regard them as sufficient to disarm the arguments presented but consider them interesting enough to warrant closer inspection of them. In the end, they will give us occasion to review the reasoning of the previous section in more detail and to bring the arguments up to date. It will then appear, I hope, that they are false.

The general trend of our discourse will thus be similar to the more fervent plea made by the sociologist, G. Lundberg, in his well known book *Can Science Save Us?*[1] He argues on very popular and somewhat general grounds for the virtues of social science and for its role as a savior from impending global disaster. His comments on the technological and, indeed, the philosophic benefits which physical science has brought and which, he believes, social science will bring are a joy to read; his insistence that modern science is among the classic achievements of all time and has the further advantage of reliability is wholesome and encouraging.

There are three points, however, with which I would take issue. One is Lundberg's statement of scientific method, which is too primitive and too empirical to satisfy the modern physicist. It is in fact similar to the "popular view" which was criticized at the very beginning of this book, but this fault is perhaps unavoidable in so popular a presentation. The second attaches to his low regard for the study of history as such, which I cannot share. As we shall see later in this book, human history, and not only special branches such as history of science, history of art, etc., is the important arena in which the norms of ethics must be tested. Finally, it seems

[1] George A. Lundberg, *Can Science Save Us?* New York: Longmans, Green and Company, 1947.

to me wrong, in the very perspective of history, to assume that science is indifferent to philosophic interpretation and vice versa. On the contrary, philosophic systems often arise from significant scientific discoveries, and the latter are strongly conditioned by the philosophic climate of their day.

We now turn to a brief consideration of instances from social science in which some features show a strong resemblance to the methods employed in the physical sciences. The first few examples have reference only to the *form* of so-called laws or rules which have been found to describe social processes. In 1913 Felix Auerbach[2] noticed an interesting regularity in the population of cities in some countries. To exemplify his findings in a crude and simple way (which does violence to present facts), we list the cities in the United States in the order of decreasing (population) size and write below each city its rank in this tabulation. Beneath the rank we record the size (in millions of inhabitants) as follows:

New York	*Chicago*	*Los Angeles*	*Philadelphia*
1	2	3	4
7.2	3.6	2.4	1.6

Evidently the product, rank times size, is a constant for all cities, and this result is claimed to be true if the tabulation is carried out to several hundred cities. The situation reminds the chemist somewhat of the abundance of elements in na-

[2] Felix Auerbach, "Das Gesetz der Bevölkerungskonzentration," *Petermann's Geographische Mitteilungen,* Vol. 59 (1913), pp. 74–76.

ture which, although limited to about one hundred in number (or several hundred if all different isotopes are listed), likewise show a relation between rank (atomic number or atomic weight) and population in the universe. In the latter instance the relation is very much more complicated than above; rank times abundance is by no means constant, but the two bear a known quantitative functional relationship to each other. The amazing thing is that the functional relation in social science is even simpler than in physical science. There are theories which account for the latter with some success in physics, deriving the abundance from certain plausible assumptions about a primordial nuclear fluid in which the elements were formed. Similar attempts by social scientists have not been equally successful, in this author's opinion, but they do take vaguely the form of physical reasoning. Nothing more is intended here than to exhibit suggestive formal analogies which, if substantially confirmed, would override the objections in the previous section.

To pursue this interesting indication a little further, we cite the contents of a paper by G. K. Zipf,[3] who proposes a generalization of the rank-size "law" in the form: $P = \dfrac{P_e}{R^p}$ In this equation P is the population of a given city within a country, P_e the population of the largest city, R the rank of the city, and p is a number, not necessarily an integer and usually near 1. By giving different values to p he manages to

[3] G. K. Zipf, *Psychological Record* 4, 1940, p. 43.

fit the cities in the United States, Germany, and India to his rule. Another article[4] offers a graph in which the theoretical relation is compared with census data over a large range of cities, and the agreement is very impressive. Something like physical lawfulness seems to be present within social science.

The literature on "social physics" is now considerable,[5] but an extended review of it is probably out of place in a book devoted to the problems of ethics. Much of it is repulsive to social scientists and philosophers, chiefly, I think, because social physics takes proved and useful physical quantities like temperature, pressure, and density into sociology and expects them to render equal service there. This is probably too simple a procedure, certainly not one required by the method of science, which demonstrates fairly convincingly that concepts useful at one level of description generally break down or require modification at another. The concepts of thermodynamics are meaningless when applied to the motion of single molecules, for the latter have no temperature, no pressure, no density. And so one should certainly expect that the concepts of thermodynamics cannot be taken over as such into social science. But, we repeat, the typical form of physical scientific laws employing concepts native in sociology is certainly present in the behavioral discipline.

[4] G. K. Zipf, *Am. Soc. Review* 12, 1947, p. 634.
[5] See, for example, J. Q. Stewart and W. Warntz, *Am. Geographical Soc.* XLVIII, 2, 1958, p. 167.

We offer another example to support this view. Several writers, among them again Professor Zipf,[6] have featured what they call the $\frac{P_1P_2}{D}$ hypothesis. It is contended that the rate of movement of persons between two communities D miles apart and having populations P_1 and P_2, is proportional to $\frac{P_1P_2}{D}$. As shown in the last paper cited, this relationship holds well for truck and automobile traffic on highways, as well as for rail and air traffic. Zipf even tried to derive this hypothesis from the rank-size law; at any rate, he established an interesting connection between the two.

During the last war, under the auspices of the United States Armed Forces, experiments were conducted on the speed of propagation of rumors. Leaflets carrying news items were dropped from a plane at a certain locality, and then, after a period time, investigators were sent to interrogate people in places at different distances from the "source of rumors." The number who had heard the news was proportional to the "source strength," (the number of leaflets dropped) divided by the distance of the place from the source. The physicist will be pardoned if he sees here a close resemblance with Newton's law of gravitation or Coulomb's law of attraction between electric charges. The same analog is suggested by the P_1P_2/D-hypothesis. The fact that the Newton-Coulomb law contains the square of D instead of D

[6] G. K. Zipf, *Am. Soc. Review 11*, 1946, p. 677.

does not destroy, but rather strengthens the parallelism; for it obviously suggests that when rumors spread in three dimensions instead of two, and when people travel in space, D^2 will replace the D in the sociological formulas.

Contacts between physical and social sciences are visible not only in the common mathematical features just discussed, but also in the dominant style of unformalized sociological reasoning. Among the classics of sociology is Durkheim's theory of suicide[7] which will be selected here for a brief review. For further details on methodological aspects of the theory, the reader is referred to Sorokin's excellent and timely analysis of it.[8] The present account leans heavily on Sorokin's paper.

The first step in Durkheim's treatment, as in every problem of physical science, is to classify and organize the empirical materal. In the terms introduced in chapter I the P-domain here consists of all factual, observational data on suicides. As in the physical sciences, these data are first classified, for data which look alike often can not be treated by a uniform theory but require careful inspection and discrimination before causes are assigned. The classification Durkheim makes distinguishes three types of suicide: altruistic, egoistic, and anomic, and each of these invites special investigation. Altruistic suicide is a sacrificial act wherein a person gives

---

[7] E. Durkheim, *Suicide: A Study in Sociology,* tr. by John A. Spalding and George Simpson, Glencoe, Illinois: Free Press, 1951.

[8] P. A. Sorokin, "How Are Sociological Theories Conceived, Developed, and Validated?", *Social Science,* 35 (1960), p. 78.

his life for the welfare of a group of which he is a member, exemplified by the hara-kiri of the Japanese Samurai, kamikaze, the suttee of Hindu widows, the suicide of an officer who violated the honor code. Egoistic suicide is the exit of a lonely individual who cannot endure the trials of life; his separation from a society that fails to support him is the cause of his ultimate despair. The third type, anomic suicide, occurs when sudden changes of environment upset a man's adjustment and composure, tear him loose from principles that previously formed anchors in his life, as in revolutions, economic depressions and the like. So much for the observational, or P-domain.

The set of constructs connected with the observed P-facts, is simple and is indeed suggested by this classification. In chronological fact it probably preceded the classification in Durkheim's mind. It involves the firmness of integration in the first class, the lack of it in the second class of suicides. Anomic suicide requires a different set of concepts of its own.

In terms of these constructs Durkheim forms a theory which is at once the postulate of his formalism—there being at present, it seems, no logically anterior social concepts from which the idea of integration can be derived. The postulate is simple: it relates a construct, frequency of suicides, to another, firmness of integration within society.

The last stage of Durkheim's work is devoted to the confirmation of the hypotheses, and this confirmation is held to be successful by workers in the field. To illustrate this success Sorokin lists thirteen items of positive prediction, among them the following in the class of egoistic suicides: the higher rate of suicide among the single than among the married, the higher rate among single than among divorced; the distribution among occupations; the high rate among vagabonds; the increase in the suicide rate in Europe during the nineteenth and twentieth centuries; even the daily and annual variations in the rate of suicides can be explained.

There are imperfections in the theory, of course. One would like to see a quantitative measure of such concepts as firmness in integration, and a numerical relation between it and the suicide rate. On the whole, however, all the characteristics of the scientific method are present and every hope for refinement is sustained by the facts. Here as elsewhere, predictions are statistical in nature, no attempt being made to assign significance to individual cases. That tendency, as we shall see, is common not merely to all successful sociological theories but to most new physical theories as well.

Enough of such particulars which look like kernels of scientific lawfulness in a largely chaotic medium, that of *de facto* human behavior. But even the domain of the normative disciplines, ethics and esthetics, has been invaded by mathematical and physical methodology. This happened first, perhaps, at least in the Western World, in the days of Socrates and Plato when they taught that virtue is knowledge. Spinoza's ethical system imitates the axiomatic structure of geometry and in some respects foreshadows a formalized system like modern relativity. Kant looked upon "the moral law within me" as the natural counterpart of "the starry heavens above me." The success of maximum and minimum principles in mathematical physics has induced many similar formulations in ethics—one wishes to maximize pleasure, freedom, responsibility; minimize pain, poverty, disease. One of the most recent suggestions in this vein, particularly noteworthy because of its specificity and boldness, is contained in R. Bruce Lindsay's[9] thermodynamic imperative, which claims that man, in his ethical behavior, should counteract the natural tendency inherent in the second law of thermodynamics and endeavor to reduce the entropy of his environment. All of these raise problems, many of them again connected with the fact that the content of ethics is not that of physics, that wholly fresh concepts must be found if justice is to be done to a different kind of experience; but the attempts to show that there is some sort of congeniality, an affinity of structure in the two areas, physical science on the one hand, social and, if I may anticipate what follows, ethical science on the other. These indications prompt us to review the arguments discussed earlier in the present chapter.

SECOND THOUGHTS ON THE SIX ARGUMENTS

We spoke of the inherently destructive effect of scientific analysis. The facts cited on that occasion (pp. 63 are cer-

[9] R. Bruce Lindsay, *American Scientist*, 47 (1959), p. 376.

tainly true, but they raise two questions which we have not yet answered. First, is there an alternative to the analytic practices of science? Second, does its method merely destroy, degrade and violate, or does it also offer novel facilities for creation, ennoblement, and edification?

The first question need not detain us long. No feasible alternative has been suggested, save perhaps by the extreme rationalist of bygone days and the mystic who believed that the actual properties of things can be revealed by a priori contemplation and by meditation. Such views are untenable today; they contradict what has been learned from science in areas where even its antagonists grant its legitimacy. One can not spin the solid contingencies of the world out of the tenuous fabric of pure reason.

When a biologist kills an organism to study its function, he does so because there is no other way to gain this particular knowledge. When the archeologist submits a sample of his ancient specimen to a mass spectrographic test he does this because the procedure yields a better value for its age than estimates based on other methods. The theoretical physicist has worried very much about the unknown and unpredictable effects of his measurements upon the state of the elementary particles he is under compulsion to study; he has tried in many ways to free himself from the verdict of the uncertainty principle but has been forced into resignation before it. Certainly, if there were a way to observe atoms and electrons without disturbing them he would seize upon it with relief and satisfaction. When he does have a choice, as for instance when absorption spectroscopy can be used for identification of a substance in place of test tube analysis, he usually employs the less destructive procedure. The upshot is: we cannot learn the things we ought to know in science except by the methods actually in use. Hence, to question the method of science on the basis of its destructiveness is to question our need to know. No reasonable person will do this in physical science; to question our need to know in social science and in ethics is equally indefensible. Let us then accept the conclusion that science, when applied to these latter disciplines, is likely to interfere in certain ways, and possibly in wholesome ways, with their procedures, but remain open minded about the extent and the severity of the interference.

Turning now to the second question, one finds, I think, that its answer is obscured by a cobweb of prejudices. Of course science does not merely destroy. The chemist who dissolves a pitifully small portion of substance is able to synthesize all sorts of compounds, both useful and beautiful, as a result of an insignificant destructive act. The uncertainty principle has actually enhanced the creative power of physics enormously, and there is nothing degrading about the cure of diseases made possible by the killing of organisms. The humanist will be the first to point out that the attainment of excellence requires sacrifice, and this is true in science, too. But in science the sacrifice is as small as it can possibly be made, and this should recommend its use at large. Our earlier simile, according to which science secures only the corpse of truth, not living truth, is altogether wrong, and if science violates the essence of human actions no more than it does the nature of material things let us welcome it; for its end effect is not necessarily corrupting, it can be truly enriching.

Those who fear the stultifying consequences of an encroachment of science upon our lives, pointing a finger of scorn at our entertainment industries, forget that their television set has a scientific device for shutting it off; they also seem to be unaware of the fact that "TV personalities" and producers are about as far away from any scientific discipline as one can get.

Does science truly deal only with the *quantitative* aspects of things, does it ignore qualities and values? The impression that it does so arises when science is seen only in its descriptive phase, where it relates and enumerates observed facts. But we have already recognized in the first chapter that the full range of explanation extends far beyond facts, that it involves ideal constructs which, though somehow related to observations and through them to numerical magnitudes, have characteristics which are entirely different from the properties of numbers.

One of the simplest and most widely used concepts of physical science is that of a mathematical function, a construct which is related to numbers much as a line is related to points. I shall not introduce here weighty mathematical arguments to make clear that a line is more than an aggregate of points, and that in a similar respect a function is more than a table of numerical values. A table has a finite number of entries, whereas a function has a continuity of values, and therein lies the crucial difference which defeats the belief that a function is merely a numerical artifact. But modern physics has grown abstract in a way which makes even a function seem very concrete; it uses the notion of an operator which is still farther removed from the domain of descriptive numerical quantities than a function, and the supposition that an operator signifies a quantitative aspect of experience is very far fetched, is in fact true only in a most distant and uninteresting sense. In quantum theory, operators without direct relations to numbers are now the constructs which replace functions as representatives of the states of physical systems. Beyond functions and operators, there are the ideas of symmetry and invariance, constructs which have reference to the form of equations and do not speak of numbers at all; they are being viewed and employed as more basic and important even than the older conservation laws, which did claim that certain numerical quantities remain constant in physical processes. It is a far cry from simple numbers to symmetry and invariance, and in view of the importance of these abstract ideas, one must concede that the time-honored claim according to which physical science is limited to the direct description of numerical magnitudes, that is to say of quantities, can no longer be maintained.

Nevertheless it must be recorded that concepts like function, operator, invariance, which are not themselves quantitative—and for that reason might be called qualitative—are *ultimately* connected with numbers. The connecting pathway, however, may be very devious; it usually goes by way of many intermediate constructs which stand between the abstract ideas and the measuring process. The fact that qualitative[10] concepts must ultimately make contact with quantitative data, that is to say with measured quantities, is a consequence of a requirement, noted in Chapter I, which insists upon the testability of scientific prediction by primary experience, a requirement which is never relinquished. But primary experience, when most articulate, speaks in quantitative terms.[11] Summarizing, we say that science in its creative theoretical phase is rarely bound to numbers, is not quantitative in that sense. But no matter how abstract its

---

[10] Here in the sense of non-numerical.

[11] To be sure, this statement needs a slight rephrasing to be accurate; it should read: The language of primary experience can be *converted* (by the intervention of constructs, to be sure) into numerically quantitative terms. For instance, science very often deals with matters of existence and non-existence, occurrence and non-occurrence, with questions which have the answers "yes" and "no." On the face of it, these questions seem qualitative, but they are not; for "yes" and "no" can always be translated into the numbers 1 and 0 and for mathematical purposes they often are thus translated. In a similar way, the qualitative chaos observed in a series of random happenings can be translated into quantitative probabilities. It is with this understanding, which permits such translation, that science must make ultimate contact with data in quantitative terms.

flights of reason, it always keeps in sight the data of quantitative experience which arise in observations on or near the P-plane. If immediate experience could not be described in terms of numbers, the gears of science could not be engaged.

This leads us to the question: Do human relations permit expression, at the datal stage, in quantitative terms? Sociologists argue that they do.[12] I shall content myself with showing that in physical science, progress is associated with the discovery of ways of quantifying primary (P) experience, leaving the reader to judge for himself whether progress in social science may not likewise accompany successful attempts at quantification.

On page 66 we drew the provisional inference that visual color was resistant to the scientist's attempt at quantification; it seemed to be a true quality which defied description in terms of wavelength and intensity. I now record the end of that story. The professor of painting was right thirty years ago. Since then, however, the physicist learned that a third "quantity" besides wavelength and intensity, namely, saturation, is necessary to define color completely. Two colors which agree with respect to all three of these variables also look alike: Nothing now slips through the scientific net, and the obscure quality color does in fact yield to quantitative scientific treatment. If this is said to be the result of a trans-

[12] See Lundberg, *loc. cit.*, also S. C. Dodd, "The Reactants Models A^a"; preprint of a chapter in a volume, now in preparation, entitled *Revere Studies on Message Diffusion*. I express my indebtedness to Professor Dodd for allowing me to see this article prior to publication.

mutation of a quality color into the quantity color, then the history of science can be shown to be full of such instances of transformation and its major successes seem to be signalized by them.

It is safe to conclude, in answer to the claim of this second argument, that science at its best always converts "qualities" into quantities during the first stage of its progress, then transforms these quantities into higher-order constructs (symmetries, etc.) which again have some of the aspects of quality but give it command of formal facilities far richer than the domain of numerical quantities could offer. And this initial conversion has always been potentially felicitous for the art which previously may have prided itself on its competence to deal with the "quality." It is indeed well known that, in the instance at hand, superior methods of reproducing paintings of old masters as well as wholly new media and creative forms have been the practical results of our scientific understanding of saturation.[13] There is nothing inherent in the method of science which makes it stop or makes it desirable that is should stop before qualities, even those qualities that are said to be present in human relations.

There can be no question about the complexity of living and, in particular, of social phenomena. They are complex in both senses of the word, excessive in the number of variables

[13] Malevich's famous *White on White,* though painted before this scientific knowledge was fully available, subtly uses degrees of saturation.

and blind as to what they are. But the *complexity* is partly *perplexity;* the state of the art is such that we do not know where to begin in enumerating variables. Hence we cannot estimate their number. But here physical science again teaches an interesting lesson. Complexity in the sense of need for an excessive number of variables *decreases* with the growth of research.

Consider once more the elementary example of a freely falling body. At the time of Aristotle, prediction of the position of a projected stone was possible (in principle, and even then not with very great precision) when the following variables were known at the beginning of the motion: Starting point, starting velocity, end point of the motion ("final cause") and composition of the stone. The last item was needed because the elements, earth, water, air and fire, had different natural motions. Today only the first two are required for accurate prediction. The miracle of simplification came about by virtue of Galileo's discovery of the crucial importance of acceleration, a very abstruse concept in Galileo's day but of great synthetic power. Similar examples could be cited in considerable number. Who is to say, then, that simplicity will forever be absent in the living and the social realm, when all of science suggests that simplicity emerges with the growth of understanding?

Control of variables is likewise a function of scientific knowledge. This is so manifest that the point need not be pressed. In further refutation of the fourth argument (page 68), however, we emphasize its substantial irrelevance: Control is not always necessary for the success of the scientific enterprise. The most highly developed physical science is astronomy, which now features simplicity, to be sure, but is quite unable to control its variables. The astronomer cannot push his stars around, he cannot assemble simpler configurations than the sky presents, cannot adjust the temperatures of stellar bodies. He has to make his observations when nature permits, and this constraint has not thwarted his efforts in a serious way.

We now return to the fifth point of our list of objections, the one involving arbitrary decisions.

This is not the place to argue the pros and cons of free will. The argument under review affirms that men make free decisions and I shall grant this point, for if it were denied the argument would lose its force and the present discussion would be unnecessary. The question is whether, if men do make free decisions, an explanation of their actions must relinquish the approach of science.

The answer, very bluntly stated, is as follows: Inanimate objects make decisions too, and science has not lost its competence to handle them. This is a shocking statement and clearly untrue if decision is taken to be a deliberate, conscious act. In that sense unconscious objects do not make decisions, and we do not wish to engage here in speculations regarding the consciousness or unconsciousness of physical objects. Let us rather deal with the question of decision in a strictly behavioristic way and admit that a decision, when viewed externally, is either an unpredictable alternative actually taken by an entity at a fork in the path of possible events, or an unexpected departure from a predicted course of events. A ball falling squarely upon a knife edge (with its center vertically above the edge) and then toppling off to the left must be said to have made a decision if this interpretation is to be adopted. The reason why this contradicts our belief is that we have good reasons to assume the existence of unknown factors in the situation which, if they were known, would have permitted us to predict the fall to the left. Again, were an object, moving along a calculated trajectory, suddenly to change its motion without any assignable cause, an idea rather close to the concept of decision would doubtless intrude itself. We regard this as nonsense because such sudden aberrations never happen in the physical macrocosm.

They do occur, however, in the atomic microcosm. No cause has ever been found for the actual outcome of a measurement upon an atomic particle when its state of motion beforehand was exactly known, or as exactly known as the laws of nature permit. What happens in a single observation is in principle unpredictable. The state of affairs is precisely that of a ball falling on a knife edge, but we are now *assured* that a search for causes is in vain. The assurance comes from the quantum theory in its usual interpretation, which we accept.[14] Evidently, then, there already exists a branch of physical science fully able to operate in the face of phenomena involving decisions in the behavioristic sense. Hence it is incorrect to suppose that science becomes powerless before the kind of unpredictability that inheres in freedom.

The indeterminism of atomic events has, however, left its imprint upon scientific explanation. It has enforced a retreat—or is it an advance?—from the use of dynamic variables which are strictly determinate to the use of random variables and probabilities. As a result, science at its best can no longer predict what will happen upon a given unique occasion; it must speak in the language of odds, of relative frequencies and long-term averages. Like Durkheim's theory of suicide, life insurance statistics and economic analysis, physics now often contents itself with computing how many individuals are involved in a certain act at a given time, it can not always point to particular individuals. From the point of view of the unity of the sciences one must greet this development as a fortunate convergence of methods, even though it seems to the uninitiated like a surrender of some-

[14] There are those who question this interpretation on philosophic grounds and wish to invoke hidden parameters in an effort to avoid the necessity of ascribing "decisions" to atomic objects. While I see no evidence in favor of this view, let us suppose for a moment that they are right. The present form of quantum mechanics will then be changed; it will revert in some way to the premises of classical mechanics, albeit with considerable complications. What matters here, however, is that even the present indeterministic form of quantum mechanics, whether right or wrong, succeeds in dealing with atomic particles admirably, and without losing its character as a science.

thing unique on the part of physics. Actually, the shift from dynamics to statistics has greatly increased the power and the range of applicability of physical theory.

Yet, when all this is admitted, there doubtless remains a difference between physics and social science in the thinking of most philosophers. When the state of a physical system is known all averages are predetermined. One can calculate the mean value (and in fact the dispersion and all higher moments) of all variables that can possibly be observed, although individual values are beyond prediction. That is to say, there is still a high degree of statistical determination. On the other hand, the actions, the free decisions of a person seem to be indeterminate even with respect to averages; freedom, it appears, implies that nothing can be said about the aggregate of human behavior. If human actions are controlled by laws like those of quantum physics, ethics will be a stochastic affair in which the relative number of good acts can be predicted for a specified individual at a specified time.

Such a conclusion may be inevitable. A reasonable denial of it will, at any rate, require more knowledge of psychology and sociology than I can muster. On further thought, however, the conclusion is perhaps not entirely absurd or uncongenial. We do speak of a person's moral *character,* and it seems that we mean by it something like the probability distribution of good acts among many acts of various kinds which this person will perform when occasions arise, or the relative frequency of his good deeds in the past. This character, this disposition, or probability can change as time goes on, in ethics as well as in physics.

Closer inspection of the argument based on decision or freedom (page 69) therefore reveals affinities between established science and the unsolved problems of human behavior which this fifth argument ignored. That argument poses an interesting problem for the method of science but does not prove it to be inapplicable to human affairs.

Almost parallel considerations rule out the transaction difficulty. The indeterminacy of singular atomic events is often viewed as interference between a measuring instrument and the micro-object, as feedback between the knower and the known. Whatever the proper philosophic explanation may be, statistical analysis manages to take indeterminacy in its stride while at the same time maintaining the full rigor of the scientific method. Better still, reliance on probabilities rather than on absolute knowledge of vagrant point events, *restores* the integrity of science which was threatened by an unmanageable complexity that appeared when classical deterministic reasoning was introduced in the atomic domain. The new style of physics, the statistical outlook it shares with sociology and economics, make unknowable things unimportant, put the accent on the lawful features of experience, where it belongs.

What, then, has been achieved by our rejection of the various claims composing the thesis that the method of current physical science is not applicable to human affairs? As previously outlined, such rejection, if valid, does not *demonstrate* applicability; but it opens the gate for such excursions as are taken in the later chapters of this book.

Let us not forget, however, what significant changes the discovery of feedback and uncertainty has brought about in physical methodology. Physics had to abandon its old mechanistic, single-event predictions and content itself with the forecasting of averages in ensembles. We expect, therefore, that a science of human behavior will likewise be restricted to a probabilistic style of analysis and prediction, quite in conformity with its present state. And in ethics, the accent must surely be on collective behavior.

This will have an important bearing on the nature of confirmation to be practiced in connection with ethical norms, disposing us at once against an acceptance of single instances of behavior as corroborating moral imperatives. We return to these points in Chapter VIII.

Résumé on Values

ABSTRACT

The diversity of meanings expressed by the word value is bewildering. In the present chapter we examine the value of ordinary things like commodities, of intangibles like health, happiness, and friendship, and of the good qualities of human beings. The latter are seen to reside, not in static properties, whether natural or non-natural, but in courses of action under freedom. These form the central business of ethics.

Questions then arise concerning the normative force of human judgments, and how it derives from the protocol data of a given discipline. In science, there are certain normative aspects connected with the idea of validity or correctness, in ethics they concern the "ought" of actions. The two are not the same, but have certain formal attributes in common. Yet the "ought" of ethics never follows logically from the "is" of science, nor from the theories of science, nor from the facts of human behavior. Consideration of the whole structure of ethical methodology, to be undertaken in Chapter IV, is necessary for establishing the ethical good.

In conclusion, systematizations of value theory, chiefly a recent one by R. S. Hartman and another by T. E. Hill, are reviewed. We leave our survey with a strong doubt about the suitability of the value concept, which is encumbered by a welter of incongruities, to form the basis for a satisfactory analysis of ethics.

PRELIMINARIES

Traditionally, the center of attention among moral philosophers is the concept of *value*. Human actions are said to strive toward the acquisition or attainment or realization of values, and if values can only be classified according to reliable principles or measured on some universal scale, then the moral quality of an action is thought to be determined by the class or the scale position of the value which the action intends to realize. This sequence of steps is at once suggested as a possibility, albeit without clear promise of success, by the simpler natural science, where temporal order is established by the position of hands on a dial, force by pointers on a scale, temperature by the height of a fluid column.

The last example is perhaps especially appropriate, for temperature regulates the flow of heat in a manner somewhat analogous to the way in which values control human actions. Moreover temperature in the subjective P-sense, like value, is not directly measurable as we have seen in Chapter I; hence a further, intrinsically unrelated quality must be selected to provide the scale. That is to say, one must set up a rule of correspondence in the form of an operational definition of value. Thus in the case of temperature the length of a mercury column, which is empirically found to vary in conjunction with subjectively sensed hotness, is called upon to serve as scale. In value theory, the scale is sometimes taken to be the pleasure associated with completed action, or self-fulfillment, or self-denial, or some other state of affairs accessible to direct awareness and judged to be measurable or at any rate discernible with sufficient exactness to serve this purpose.

It is only fair to notice, however, that while this procedure works as simply as is here predicated in many instances, science also presents numerous examples in which the connection between a regulative property (like sensed temperature) and the scale of measurement is far more remote, where the scale is non-linear or even multi-dimensional, and others

RÉSUMÉ ON VALUES

where a scale has not been discovered at all. To name a few such instances for closer consideration, one might select entropy (remote rules of correspondence), psychological response (non-linear function of stimulus intensity), and national income (no objective scale at all).

The rule of correspondence between directly experienced value and its measure or its epistemic correlate may, therefore, be more complex than value philosophers realize.[1]

As to the thesis that human actions refer to a scale of values and that values are measurably correlated with some property produced by or accompanying actions, science offers neither support nor refutation, even if it suggests the possibility by analogy. Strictly, then, the thesis must be subjected to logical scrutiny and to the test of factual success. It could fail because the concept value, even in an immediate sense, might not be capable of sufficiently precise definition to be referrable to any unique scale; or values may not be the purposes (or the causes) of human action; or the chosen scale itself might prove unreliable. It is my belief that all these misadventures have occurred and are secretly troubling those who travel into ethics through the theory of values; that the concept of value itself is at present the object of a

---

[1] Numerous ways of measuring values are discussed by C. W. Churchman, *Prediction and Optional Decision*, Englewood Cliffs, New Jersey: Prentice-Hall, 1961. This book is interesting in the present context because it points up the difficulties encountered in the process. In spite of this, Churchman proposes to establish the ought via values, suggesting that right action is the action man would choose if he were completely enlightened and free!

global rescue operation involving most of the resources of moral philosophy. Worse than that, I doubt whether the operation can succeed. To be sure such violent claims as these need careful substantiation and require, above all, that we begin by taking the concept of value seriously and appraise its competence.

Another kind of question can be asked concerning values. Quite apart from any success of the attempt to infuse action with value and then determine value by reference to some more easily discernible attribute, one needs also to make sure that this elaborate correspondence when established is sufficient to complete the task of ethics. It is conceivable, for example, that it could satisfy the scientific query as to what men do, how they in fact behave, leaving unanswered the question whether it is right or wrong that they should thus behave. In that case the theory of values might serve as a basis for cultural anthropology and for sociology, but not for a kind of ethics which asks for something akin to conceptual truth beyond the factual in science, which requires a standard of judgment higher than observed behavior. Such a standard may be illusory; but again a fair-minded approach should concede its possibility.

Here, however, even the analogy with science raises a problem of grave difficulty, and we do well to confront it at once. Value, most people believe, is the counterpart of fact; what one is to ethics the other is to science. This is especially the view of those who see no bridge between the two, who believe that ethics and science are poles apart and that the methods for exploring one area are completely different from those suited for exploring the other. It is the view which, barely concealing its envy, accounts for the practical successes of science by saying that science sticks to facts and gains power by that limitation; and it craves for ethics the same restrictive preoccupation with the contrary of fact, namely value, which will confer upon ethics a similar good fortune. Those holding this view seem unaware of the plain truth that theoretical science, the immensely successful kind of science, would never have attained its accomplishments had it confined itself only to facts. As we have endeavored to show in Chapter I, the structure of science contains far more than facts, and one rightly wonders whether the structure of a functioning ethical enterprise must not contain more than values, whatever they may be.

There is a peculiar paradox in this surmise. The person who believes in the sufficiency of facts for science is, generally speaking, the positivist. He who holds that values are sufficient for ethics usually scorns positivism as too narrow a doctrine, remaining unmindful, it appears, of the circumstance that he is asserting for ethics in effect what his opponent says for science. If a parallel can be drawn at all, then the ethical positivist is a person claiming that values are the positive components at the core of ethics, and that their understanding completes the quest of moral philosophy.

The Meanings of Value

Since the word value is called upon to perform a most important service, it may be well for us to review its linguistic origin. Somehow, our comprehension of things, situations and people is deepened, our interest in them quickened by a knowledge of their ancestral histories, and this should certainly be true for words that are employed as pack animals. In the present instance, we discover an interesting variety of meanings among the ancestors, one of which—and not the primary meaning—has come to be reflected in the present usage of the word.

The Latin *valere,* from which value is derived, meant primarily to enjoy physical strength. The adjective *validus* is frequently applied to a bull (*taurus validus*). But the word took on related meanings: to be capable in a more general sense, to be powerful, influential. The state of health was always included in its Latin connotation, as is evident in the Roman farewell: *vale.* Metaphoric diffusion goes on from powerful to fitting and appropriate and finally to the notion of equivalence of coins (*"dum pro argenteis decem aureus unus valeret"*). Here the original comes close to our present version of values, as it does in the "equivalence" of words. For in Latin, synonyms are words of equal value (*"verbum, quod idem valeat"*). Rarely, however, would the Roman have used *valere* to refer to anything of value in our sense; this would have been more accurately rendered by **pretium, honos,** or **virtus.**

RÉSUMÉ ON VALUES

If value has been upgraded, from physical strength to general worth, the word price has suffered the opposite fate. It started approximately where value ended, with the Latin *pretio,* and has now come down to its mundane usage in economics. The French *prix* and the Spanish *precio* still retain a little of the more "precious" flavor. When later I seem to be degrading values by linking them with prices, I hope the reader will remember this idiosyncrasy of our language.

We now inspect the meanings of value in today's context. Simple words implying value judgments are good, pleasant, beautiful, genuine, honorable, virtuous, and their opposites. They can be applied to several different types of noun, those denoting things, persons, and indeed processes or courses of action. For the present we restrict our discussion to things and persons. One peculiar aspect about the assignment of a value to an object, often noted in the literature and greatly emphasized by G. E. Moore,[2] is its indirectness of reference. When we say an object is good we do not mean it is blue, large, heavy, or that it possesses any other particular sensory qualities; yet in another sense we mean all of these. Goodness is not a natural property of objects, but still it expresses itself through their natural properties. For this reason, goodness and more generally value, is often termed a non-natural quality of things. Here then arises one of the much discussed problems of value theory: Are non-natural qualities

[2] G. E. Moore, *Principia Ethica,* New York: The Home University Library, Oxford University Press, 1912.

reducible to natural ones; are they a collective function of all the natural qualities of an object; do they refer to something aside from the natural qualities such as the reaction of a human mind or the use to which the object can be put; or are non-natural qualities irreducible and *sui generis?* The last proposition is affirmed by Moore, who argues profoundly and with manicured verbosity about the logical difficulties inherent in every alternative. This tendency continues, though often with different results, in the current writings of analysts and many others[3] and has occasionally led to clarification and removal of obscurities in value theory.

The present inquiry is meant to move in a different direction, for we are less interested in the logic of usages than in the dynamics of ethical procedures. For this reason we abandon the study of values as non-natural qualities of things. I should be less than frank, however, were I not to record a modest note of disapproval with respect to the premise of Moore's arguments. There is, I think, no useful or even plausible way of distinguishing natural qualities from others. If natural means *directly perceived,* then beauty, loveliness, and perfection are often the most natural qualities one can apprehend. If natural means *physical* and adverts to properties which a scientist can quantify and measure, it includes many *functional* ones, and functional is evidently the char-

[3] See, for example, R. M. Hare, *The Language of Morals,* New York: Oxford University Press, 1952; P. B. Rice, *On the Knowledge of Good and Evil,* New York: Random House, 1955.

acter of those aspects which the term non-natural is meant to cover. Energy, for example, is functional; it is not directly exhibited but represents, in textbook language, the amount of work an object is able to deliver. He would thus have to conclude, paradoxically, that natural properties include non-natural ones. Some would say, to be sure, non-natural does not mean functional either, and the chase for meaning would go on. But I fear that it would ultimately seize upon nothing, since I suspect that the element that distinguishes the non-natural from the functional is merely the vagueness of the term.[4]

At any rate, values are properties or attributes of things. They attach, first of all, to concrete objects. Here, value is generated by the use to which they can be put, and this use is either a limited one in the status of loan or lease, or it is unlimited in outright ownership. There are philosophers, to be sure, who would wish to claim that it is meaningful to speak of the moon as good or bad. Any sense in which such a statement is interesting prior to the day when the moon serves human purposes will here be disregarded. When objects are usable they are commodities, and their value depends in a complicated way on our desire to have them and on their availability, on what is loosely called the law of supply and demand. The complexity of this relation transcends, it

[4] For similar criticisms, see C. D. Broad, "Moore's Ethical Doctrines," in the *Philosophy of G. E. Moore,* ed. P. A. Schilpp, Evanston and Chicago: Northwestern University, 1942.

seems, all present efforts to formalize it; in particular, the market price is not necessarily an index of value, as is easily seen in the instance of the water we drink and the air we breathe. Here, the desire for use, the demand, is an unconscious need and the availability is practically unlimited; hence, the value, sometimes called intrinsic value because it refuses to be fixed by formulas, is quite unrelated to the price we pay.

Values are also attributed to intangibles like health, pleasure, happiness, friendship, security, and leisure. Different principles begin to operate here, primarily because one of the components which determine the value of commodities ceases to be significant: these intangibles can be created at will, they are not subject to laws of material supply and hence indefinitely available, resembling in this respect the water and air just cited. To the complexities aforementioned there is now added another, very fundamental one, namely the observation that these intangible goods defy the laws of arithmetic, which are applicable only to discrete things. Hence we have no calculus suitable to an assignment of value in these instances.[5] They are, at least at present, unmeasurable and are for that reason often called *qualities* in contradistinction to *quantities*. (Cf. Chapter II.) However,

[5] For an interesting attempt to formulate such a calculus, see N. M. Smith, Jr., "A Calculus for Ethics: A Theory of the Structure of Value" (in two parts), *Behavioral Science*, Vol. 1, Nos. 2 and 3, April and July, 1956. This is the most serious and circumspect approach to value theory in terms of operations research which has come to this author's attention; it clearly merits study by philosophers.

let us not conclude that this admission places intangibles beyond the pale of principled treatment or of legitimate ethical concerns. Nobody will contend today that only measurable magnitudes, that is to say quantities relatable to a linear continuum, are important. As we have seen on two earlier occasions, even natural science invokes constructs like matrices, groups, and fields which have none of these simple attributes; and modern mathematics soars far above the simple linear manifold, as we have also noted, indeed far above the strictly measurable, as in topology. Hence, it is by no means certain that we shall always remain unable to deal mathematically with intangibles, unable to develop a calculus for them which will enable us to specify and formalize their value. That time, however, is not yet.

Meanwhile, then, cruder means in the form of intuitive devices are forced upon us in the endeavor to appraise health, love, and pleasure. One possibility is to select arbitrarily one of them, say pleasure, as elementary, and then adjudge the competence of others, such as health and love to induce the elementary good. I used the word "arbitrarily" with full cognizance of its disturbing effect upon my argument, for I know of no rigorous criterion that would accent one "good" more heavily than another. Attempts go forth in many places to show that there is indeed an elementary or a primary good, related by different writers to different terms such as human self-fulfillment, or the goal of human evolution or the survival of the race. All of these basic principles, it seems, must beg the final question as to why they should be regarded as good at all. No amount of scientific erudition, no logical argument can place one of these above the others. It seems to be up to a discipline like ethics to make a *choice*.

It is utterly astonishing to see how many moral philosophers, though rigorously trained in logic, and often exclusively in logic, persistently balk at the *need for choice* for commitment which confronts us here. One of the deepest insights conveyed to us by this branch of modern philosophy concerns the impossibility of establishing a formal system that is assuredly self-consistent and complete, and certainly none that can do without *chosen axioms* and *primitives*. It should be obvious, therefore, that at some place one must meet a legitimate occasion for choice, and apparently we have met one here. Let us face it, then, without embarrassment and acknowledge that the selection of the principle in terms of which one decides to measure value is not regulated by any non-moral discipline and depends on human choice. Different choices yield different systems of value, according to the present (incomplete) analysis. Later we shall show, however, that the selection of a principle here under consideration, while essential, is not sufficient to establish values.

At this point we summarize by saying that the value of intangibles is estimated intuitively by the degree to which they contribute to the creation or maintenance of some primitively defined "absolute" or elementary good. We have spoken of intangibles as *possessing* value. Sometimes happi-

RÉSUMÉ ON VALUES

ness, health, friendship and so on are said to *be* values. The difference is usually semantic and requires no further elaboration, except on a very deep plane of ontology. If the question is whether values truly exist, then a distinction between being and its attributes may become important, for if happiness exists and value is its attribute, value can hardly be given specific being of its own. To say that happiness *is* a value reflects, therefore, a Platonic premise, which to some minds is an asset. This book, in ignoring ontological aspects of value theory and ethics for lack of space (and of the author's competence), does not wish to detract from their possible relevance.

The preceding paragraphs dealt first with the value of common objects, limited and unlimited in their availability, and secondly with the value of intangibles. At the next stage we encounter living things and we ask about their value. Life itself, of course, has value, but it belongs to the former class of instances. One might attempt to make the transition from there to the value of living things by imparting to the latter simply the value of life, but this clearly fails, for not all forms of life have equal value. Similarly, the addition of other intangibles beside life does not suffice to yield a measure of value: when one speaks of a good person his goodness is certainly not meant to be a function of his perceptible qualities or even the intangibles with which he is endowed, since they can be used for good or evil. A happy mood does not make him good, neither will health, nor love of his parents or his fellowmen. Life and all the rest of these qualities do not constitute his goodness even though they may be a necessary condition for it. No, the goodness of a person is evidently dependent on what he *does*. An important change is thus seen to occur as we pass from the study of value in inanimate objects or in abstract qualities to the domain of the living, where "free" decision is possible together with voluntary action. Here the meaning of good, and indeed of the entire value concept, is transferred from its lodgings in external qualities—where it resides for things—to the internal dynamics of willing and acting. A person is good because what he does is in accord with certain rules.

In the shift to the living, then, we have made the transition to ethics. An egg that is good and a boy that is good are incomparably different in the connotations of good. It is difficult to overemphasize this point. Precisely because it is sometimes ignored, we witness wholly ineffectual efforts at rationalizing ethics. For instance, a thing is said to be good or valuable, if its conforms to its own essence. If essence is properly defined on the basis of experience, this statement may be meaningful. But if the phrase is then extended to read: man is good if he conforms to the essence of man, the extension involves a double infelicity. In the first place, while the essence of an egg may be ascertainable without great difficulty, though always with some ambiguity and relativity of meaning, the essence of man is altogether beyond factual discernment, as innumerable studies by cultural anthropologists

RÉSUMÉ ON VALUES

have shown. Apart from this, however, the claim denies the crucial element which separates man (and other animals) from inert objects, his freedom, and his capacity to act. Man is good because of his actions, the egg because of its inert properties.

Having encountered and recognized what constitutes value in man, we have likewise uncovered the main springs of ethics: Will and action in accordance with certain rules. The remainder of our discourse on ethics will mostly relate to these rules and to the manner in which actions can accord with them. It will explore their nature and their validity if they have validity, their origin and their suasive force. At present, however, we return to the problem of values, both those residing in external qualities and those inherent in human actions.

Methods for Assessing Value; the Price of Things

To take the concept of value seriously one must certainly attend to the concrete ways in which it has been, or can be, measured. The measure of value, at least among simple things, is *price*. To be sure, price need not be expressed in money terms, but wherever it is thus expressible the situation is clear and easily tractable. Hence we review first the chief methods for assigning prices.

Even in economics a distinction is made which reflects the contrast between the value of inert things and the action value of the living; for we contrast the value or the price of goods with that of services. Although ethical considerations do not enter here, the dichotomy at once appears, and it has indeed dominated economic thinking to the extent of producing two incompatible types of economic theories of value: those which reduce the value of services to the value of goods and those which invert the reduction. The former appear to dominate capitalistic thinking, the latter, called labor theories, were initiated by the work of Ricardo, Adam Smith, and Marx, and have come to be part of the creed of communism.

The labor theory of price is simple in principle, for its essence is that the price or value of a commodity is proportional to the amount of labor needed to produce it. Amount of labor means man hours of work, and capital is merely stored-up labor. The difficulties clinging to this theory are evident and well known. They include its inability to define the price of things which are desired and used but are not man-made, its incompetence to deal adequately with rents and interest. And above all, it seems to raise as many problems in connection with the meaning of the term, amount or quantity of labor, as the use of that standard allows it to solve. For clearly a simple reckoning of man hours cannot form a proper index for a mason's labor unless it is prescribed how many bricks are to be laid per hour. Then there is the variability of skills and of degrees of training, which leads to the question whether the work hour of one man is equivalent to that of another. In spite of these difficulties, however, the

RÉSUMÉ ON VALUES

labor theories are notable because of the honest and open way in which they seek a simple measure of value. They are no more reprehensible, and also no more successful, than philosophies which seek to identify value with pleasure, love, or self-fulfillment.

When labor is disregarded, prices must be established by reference to the procedures of the market place. Since every sale is fundamentally an exchange of goods, the exchange value or exchange price became one of the earliest indices of value, sometimes called the power of exchange. It is a factual and fluctuating measure, expressible in the first instance only as an infinite number of ratios, one for each pair of commodities, specifying how much of one a customer would exchange for another. The price structure would be the set of established ratios, the number of eggs per loaf of bread, gallons of wine per suit of clothes, hours of work per automobile, etc. To convert these ratios of pairs into simple numbers one chooses a basic commodity, called money, and uses it as the denominator in every ratio. This procedure yields factual prices which reflect the preferences and habits of each consumer. They are, therefore, subjective, and they fluctuate in response to changing appetites, preferences, and habits. The merchant must guess these prices, secure goods for less and sell them at a profit.

The theory just sketched has the virtue of realism. Lacking sophistication, it remains in the subjective realm and is powerless to satisfy any desire for stability of values. It is purely descriptive, stalks behind events and imputes values as afterthoughts. As might be expected, attempts have been made to improve matters in some of these respects, to make the exchange theory of value less subjective and variable. One of these invokes the concept of "marginal utility" and makes it, in part at least, the determinant of value. The utility of a good is not necessarily its usefulness in terms of some objective standard but its importance to consumers. The principle of diminishing utility affirms that progressively diminishing importance is attached by a customer to successive increments of a purchasable commodity: If in time I buy 4 suits of clothes, the first from need, the fourth to satisfy a desire to be well dressed; the first suit was obviously more valuable to me than the third or fourth. Marginal utility is the utility, the satisfaction, conveyed by the last increment of a commodity a consumer thinks worth acquiring. It is this marginal utility, according to the refined exchange theory under discussion, which determines the subjective price of the commodity, and the market price is supposed to be proportional to it.

This may suffice here as an illustration of efforts designed to improve the simpler method of measuring values. What they achieve in securing or confirming the concept of value, or in making it more explicit, is difficult to say. They are expressions of human desire for concreteness and precision, for objective appraisal of meanings, and they should be appropriately credited as such. Nor is it reasonable for us to limit our endeavors at concrete assessment of value in terms of price to commodities and routine services. What is discouraging in the ideas reviewed above is not the goal they seek but only their lack of success in attaining it so far.

It is in this sense that we should also consider the pricing of intangibles, although the instinctive deprecation conveyed by this phrase is strong. Somehow, one feels that quantitative appraisal is foreign, degrading and violating to the higher values and one should desist from it. Yet there is ample evidence that some higher values have not suffered or become degraded by quantification—such values as the beauty of design which can to some extent be quantified by symmetry characters, intelligence which can be measured (though perhaps with some reservations as to significance), health which can be certified to the satisfaction of insurance companies. Yet because of the accustomed association between price and commodity we think it shocking or humorous to find money values affixed to intangibles, as did the author when, in signing a payment voucher for a lecture given in an industrial laboratory which insisted on compensating him for his effort, he noticed that the "goods" furnished were described as "consumed while being delivered."

We do price intangibles, sometimes in terms of commercial coin. Nobody thinks it improper to pay money for admission to a theater or a concert hall, institutions which by providing entertainment dispense pleasure; neatness of appearance is purchased in barbershops and education in schools; even life is sometimes valuated in terms of money, for example, when a firm insures its officers. One may not wish to regard this as a direct estimation of the worth of life, but rather as a measure springing from prudential considerations. There are, however, more drastic situations where this generous interpretation will hardly do. The life of a soldier in battle is in fact appraised in terms of material or strategic objectives, sometimes in the money equivalent of the training he has received. It was a painful experience for me to read in a military report[6] issued after the Second World War that the life of a fully trained soldier under specified circumstances was $200,000 and had to be counted as such in strategic computations of risks; nonetheless it was evident that a figure of this sort was useful to the general in making his decisions. Admittedly, these are abnormal cases which may have little importance in a philosophy of values.

The vulgarization of intangibles, if that is the true description of tendencies which aim to express the value of intangibles in terms of currency—and I think it is—goes on with increased vigor in our public life. Our advertising agencies flourish because of it, for they have discovered not only ways of assigning money value to abstract qualities, but also methods for inflicting their tastes upon the public without causing revulsion. The scholar used to be a man of learning, pride, and integrity. He still has some of these qualities in a

---

[6] See also N. M. Smith, Jr., S. S. Walters, F. C. Brooks and D. H. Blackwell, "The Theory of Value and the Science of Decision," *Journal of the Opinion Research Society of America*, 1 (1953), p. 103.

RÉSUMÉ ON VALUES

sense that is becoming extinct, for what sets him apart from the man in business and the man of affairs, above everything else, is that he has not discerned the power of money in securing reputations. He still toils painstakingly, modestly, inefficiently to acquire a "name," while the man in business pays money to have his name splashed over the countryside. An appliance dealer recently ran a newspaper ad which displayed his name topped with a crown; the ad went on to make extravagant claims regarding his honesty and human kindness. What matters here is the established occurrence of such practices in one part of the social world, an apparent unawareness of its successes in another, especially in the world of scholarships. Indeed, one can buy intangibles for money.

That, however, is not the only method of valuating them. There are other prices, sometimes spelled prizes. Special qualities are publicly rewarded in *symbolic* fashion; a correlation is set up not between intangibles and tangibles like money, but between intangibles and certain semitangibles. The titular structure of some societies is a case in point. Here the degree of excellence of persons, their value, becomes outwardly manifest in the names they carry. The very designations of trades and professions, the hierarchy of academic and military ranks, the classification of jobs in industry, the assignment of managerial responsibility, the august intonation of the word "executive," all these bespeak the propensity of our own society to value intangibles in unmistakable ways.

In certain sense these procedures also place premiums on

courses of action and are therefore fraught with ethical implications, since men receiving these awards have demonstrated their merits by their previous acts. The method of empirical pricing is not limited to commodities and static qualities of perfection, but it seems to work, within limits, for human actions as well. Still the question remains: Is this method, or any of its variants, sufficiently direct, suggestive, and conclusive to form a solid basis for ethics? Before this question can be answered it is necessary to focus attention upon another aspect of values, hitherto neglected, namely their *normative* claim. The next section deals with the problems raised by this claim and with its significance for ethics. But first, let us clarify terminology. When, within a given group of people, an actual poll is taken of beliefs, attitudes, preferences and actions, the results obtained imply the *de facto* values of the group. Sociologists and anthropologists largely restrict their interest to this type of value. Most of us feel, however, that there are supervening principles of a higher sort whose application allows a further appraisal of *de facto* values, allows a judgment whether they themselves are good or bad in a loftier sense. We speak of that sense, whatever it may be, as *normative,* and of the values which conform to it as normative values.

The Scale of Oughts

Preceding parts of this chapter have dealt with the nature of values and the ways by which they are established and made manifest. Each value, however, when established, still prompts us to ask whether indeed it *ought* to be a value and, if it is, whether our assessment of it is right. A fact is what it is and must be accepted as such; its essence is fulfilled in its being, not in its being right or wrong. Values, on the other hand, have both a factual and a normative component, the latter declaring itself in the judgment of an ought.

We encounter the second component along with the first on all levels of valuation. A thing may be valued or priced in accordance with custom, subjective want, or the laws of supply and demand. Its factual value is thus fixed. Nonetheless we wonder whether the value or the price is *right*. It may be, of course, that our wondering is idle and unjustified, amounting to nothing more than *displeasure* at the price as one may experience displeasure at a fact. This is indeed what many current theories maintain, as we shall indicate. According to them, if I say the value or the price of this commodity is too high, I am expressing my personal disapproval, and I may refuse to buy it. Argument about it on objective grounds, an appeal to something other than feeling or expediency is precluded, although arguments may still arise. I may point out that I can buy the object at a lower price elsewhere, or that the seller's profit is too high; I can persuade, but not reason cogently. If this be true, what we have called the normative aspect of the value in question is illusory, or at any rate only a matter of subjective feelings without rational justification.

Those who believe, on the contrary, that it is objectively proper and meaningful in a deeper sense to ask the question of ought must be able to point to some standard other than feeling in order to add cogency to their appeal. Such a standard might be divine commands, a sovereign's decree, an enacted law, the consensus of the nation, collective happiness of a large group if this is accepted by the group, universal happiness, and many other superfactual considerations. It could be a combination of any of these.

Notice the occurrence of *collective* happiness in this enumeration. Is it consistent to regard the happiness of many—assuming that it can be defined and recognized—as an objective principle when individual happiness is subjective? It seems to me that this conclusion follows almost from the meaning of the terms; for objective in the present context can only mean intersubjective or transsubjective. But the distinction is not purely logical or semantic, it is practical and useful as well; for example, if the argument concerning price did arise, my individual feeling would carry no normative weight while the happiness of the nation would, because it transcends the singular situation in which the argument arose.

Certainly, some of the standards mentioned have greater hortative or decisive force than others, and their mention invokes different emotional responses in different people. The point is, however, that each of them provides a basis for settling value disputes. Notice, too, that some standards are restricted to national groups while others encompass humanity.

For the simple purpose of discussing the values or prices of goods the standards here assembled are needlessly elaborate and grandiose; indeed the occasion for discriminating between the factual and the normative seems quite remote. Yet the distinction occurs even on this lowest plane and calls in principle for such elaboration as we have suggested. Then, as one moves to higher planes of value, the problem of the ought presents itself with greater urgency and at the same time, as if in compensation, the area of dispute concerning the standards introducing the ought is lessened.

Little need therefore be said about the normative aspects of intangible values. Life, health, and honesty have hortative power in the overriding judgment of most men, except in the transactions between different national groups. Within a group an appeal to some explicit standard becomes necessary in instances of conflict, when life may have to be sacrificed to honesty, or honesty to life, or when one person's life demands the death of another.

It is in the vast arena of specific actions that the ought becomes most imperative. And precisely there the distinction between the factual and the normative is widely disregarded. Value in the social sciences is too frequently identified with the actual behavior, with observed preferences of people within a group, and this is then often tacitly elevated to a norm. There is classical precedent for this confusion. The lawyer (and most inept philosopher) Cicero says in *De Officiis:* "Customs are precepts in themselves." In Montaigne's *Essays* we find: "Les loix de la conscience, que nous disons naître de la nature, naissent de la coustume." Even language tends to erase the distinction between the factual and the normative by providing very similar labels which obscure the disparities: *e.g., ethos* and *ethika, mos* and *moralis, Sitte* and *Sittlichkeit,* morale and morals.

The reason for this oversimplified treatment of social situations is easy to see: preferences are observable, statistically measurable while norms are not, and the view prevails that what is measurable becomes *ipso facto* scientific, and everybody wants sociology to be a science. If the analysis of Chapter I is accepted, then it must be clear that the philosophy of science thus implied is as truncated and rudimentary as the treatment of values which it tolerates, resulting in an erosion of the deepest moral concepts. When common practice is accepted as norm, the ideas of obligation, honor, guilt, remorse, and retribution undergo a transformation which leaves them only as shallow psychological phenomena, their victims at the mercy of psychiatrists.

This happens, not because the factual findings of social scientists are accepted, but because they are accepted as complete and ultimate. No system of ethics, however lofty in its reliance on standards, may disregard Kinsey reports and crime tables; ethics must start from there and ascend to a region of principles from which the rightness of acts can be judged, and in doing so it must insist on relevance of principles to acts. The major problem, as I see it, lies in establishing such relevance, for often in the past the normative principles of the moralist have been exalted and detached, permitting no grip on the concrete actions of living men. To remedy this, we shall propose in the next chapter a union of two kinds of standards which will suffice for normative judgments and couple them with actual behavior.

The present section was entitled the scale of oughts, a phrase intended to convey, first that the normative note rings audibly in every value judgment, and second, that its intensity rises in a crescendo from mere detectability in the assignment of material values to imperative urgency in valuating human actions.

Does Science Contain Normative Elements?

Facts, we saw, have no normative properties. A fact is, and nobody can make an is into an ought. Does this leave science without normative qualities altogether? The answer is affirmative, of course, for those who believe that science is a body of facts. From Chapter I, however, it should be clear that this belief is erroneous; science contains rational elements whose status in experience is quite different from the coercive, existential, indubitable character of facts. Constructs of science are to a large extent man's own addition to the spontaneity of the factual, and their coupling with facts is not absolutely rigid.

Constructs, when combined into theories, explain the facts either correctly or incorrectly. Hence it is possible to ask the question concerning their truth, at least in the face of given factual evidence. With respect to the facts themselves there is no truth, there is presence or absence, existence or nonexistence. To be sure, a *report* of facts may be true or false, and this precisely because the report transcends the facts, associates them with words or symbols which are constructs linked to the P-counterparts by rules of correspondence. Let us examine here briefly one of the multiple meanings of truth which is applicable to scientific theories.

This version of truth is not identical with the truth of a factual report or an account of psychological experiences. Newton's laws of motion are true (within limits) because every consequence that flows from them has its counterpart in factual experience, and no single factual experience contradicts them. In addition, Newton's laws are simple, satisfying and logically coherent internally as well as in their connection with other physical laws. The reader will, I hope, forgive me for the technical inaccuracy of the claim that Newton's laws are never contradicted. We know they are at odds with well-known relativistic observations, and insofar as they suffer contradiction they are indeed not true; they are approximations to true laws whose form is not completely known. Quantum theory, for instance, is "truer" in one limiting case and relativity in another. Applied to the ideal theory, the statement is accurate. It is important, therefore, to distinguish the truth of propositions about facts from the truth of scientific theories: The former is truth of accurate registration, the latter is truth by correspondence with facts and by formal coherence. This we shall simply call scientific truth.

Obviously, then, there inheres in a true theory an element which facts do not possess, and this element might be labeled normative. Such usage is suggested, perhaps, by the observation that most languages apply words equivalent to right and wrong to theories and also to value judgments; hence, if normative means right, scientific theories can be normative. But let us watch a little more carefully what the normative ingredient of a scientific theory achieves. It allows an economic and pleasing representation of many facts which are its consequences; it enables prediction of P-experiences; it performs the function of explanation inasmuch as it subsumes factual particulars under universal propositions. It might even elicit approval and aesthetic satisfaction because of its elegant logical structure; in sum, it regularizes many confusing aspects of our cognitive experience. Yet in spite of all these successful functions of science no one ever says that the earth *ought* to attract the moon.

True scientific theories confer regularity and predictability upon cognitive experience; for this reason I should be perfectly willing to call them normative, provided, however, allowance is made for a difference in the meaning of that term when it is applied to ethical principles. The need for the difference is perfectly clear: science deals with knowledge, its business is cognitive understanding; it has no room for *obligation,* which arises only in the sphere of human actions based on conscious freedom of choice. The difference is to be acknowledged not because of an incompetence or imperfection of science or, conversely, because of the supremacy of ethics; for it arises merely from the disparity of the fields in which science and ethics operate. One is a trustworthy guide to factual experience, the other to responsible action, and insofar as passive experience is different from active experience, the normative forces of science must differ from the normative forces of ethics.

This state of affairs is often expressed by affirming that an *is* cannot be converted into an *ought,* a proposition which follows directly from everything we have said. If left unattended, however, the statement can become a divisive dogma. The important recent book by Hall[7] clearly recognizes the difference just emphasized, but it also ends with the insight, leaving science and the domain of values cleft apart. I should like to carry the inquiry one step further and examine the normative in science in the hope of finding clues for our understanding of the (admittedly different) normative elements of ethics. The reasonableness of such an effort becomes apparent when we reflect upon the genesis of the idea of "validity" in science: we have shown in Chap-

[7] E. W. Hall, *Modern Science and Human Values,* Princeton: Van Nostrand, 1956.

ter I how elaborate, delicate and unique that idea is, how fully it engages the resources of fact and reason; to the unbiased philosopher the development of this idea is certainly as surprising or, if you please, miraculous as the evolution of an "ought" from the factual values of ethics. Furthermore, the truth of science can no more be perceived if one's gaze is fixed upon facts, than obligation can be seen in feelings and preferences. In both realms, science and ethics, the philosopher's view must be enlarged to include the total structure: facts, constructs and axioms in the former, human behavior, values and, as we intend to show (Chapters III and IV), commands in the latter. But before leaving the subject of values, we wish to present in the next section a review of current writings in the field.

THEORIES OF VALUE

There are several recent publications well suited to amplify the discussion given here. One is a collection of articles under the title "New Knowledge in Human Values."[8] Though low in systematic coherence, this volume offers a variety of (partly conflicting) views held by a very international group of authors. Next is a brief review by the German authority, F. J. von Rintelen,[9] with an ontological flavor. It will complement the present version, whose emphasis is

---

[8] *New Knowledge in Human Values,* ed. by Maslow, New York: Harper's, 1958.
[9] F. J. von Rintelen, *"Wertphilosophie,"* in *Die Philosophie im 20 Jahrhundert.* Ed. F. Heinemann, Stuttgart: E. Klett Verlag, 1959.

predominantly epistemological. The same volume contains a very circumspect, less theoretical paper by F. Heinemann.[10] Special reference will here be made to a survey article by the axiologist (Greek: *axios* – value) R. S. Hartman,[11] for it proposes a useful classification into which, formally at least, different philosophical attitudes can be fitted. In presenting it I shall take the liberty of stating my reaction to those features of Hartman's scheme which seem to me inadequate, while assenting to the general approach. Two fundamental distinctions are made, one with respect to the very knowability of values, the other with respect to the place in human experience where evidence bearing on values may be found. The first separates the cognitivist from the non-cognitivist; the person who believes there are values which can be known from him who says there are no knowable values. Hartman inserts also an intermediate position characteristic of the semi-non-cognitivists, as he calls them, and assigns to them the belief that either there are really no values but there is ground for discussing the concept of value, or there are values which escape the grasp of knowledge.

This terminology bristles with problems. Especially troublesome is the existential phrase, "there are." This seems to be taken in the simple unreflective sense of the logician to whom the symbol "there exists" is clear and devoid of mystery. Values are treated like ordinary things.

[10] F. Heinemann, *ibid*.
[11] R. S. Hartman, "General Theory of Value"; in *Philosophy in the Mid-Century*, ed. Raymond Klibansky, Florence, Italy: La Nuova Italia Editrice, 1958.

RÉSUMÉ ON VALUES

Hartman's second distinction is between the empiricist, who finds the essence of value in "experience"—presumably the contingent protocol experience designated by P in the first chapter of this book—and the formalist, who looks for it in the structure of judgments or propositions. Under experience he notes another differentiation, saying that it is either natural or non-natural. The use of these terms, though common in the literature of the subject, is to me deplorable, as indicated earlier. While the meaning of natural seems clear, its opposite is very vague indeed. For if it adverts to entities or qualities not found in inanimate nature it lacks a point; if it excludes all nature, physical as well as human nature, it leaves out values. Usually, people mean by "non-natural" either "artificial" or "supernatural," and I believe neither of these is intended.

The theory of knowledge presented in Chapter I commits one to the conception of "natural" as covering all matters which lie within the competence, the range of explanation of present natural science. Its future range is unknown, but there is no intrinsic reason why it should not comprise all that is knowable, interesting or indeed of any intellectual concern to man, since the method of science is cognate with the most basic drive for understanding. In particular, formal sciences cannot escape being natural. But this is clearly not the use to which the word is put in the article under consideration. In effect, two meanings are associated with the word non-natural. One refers to what we have called the normative aspect of value, that which sets it off so markedly

ETHICS AND SCIENCE

from fact; the other suggests inapplicability of the laws of ordinary logic to value phenomena and tries to invent a calculus appropriate for them.

After this introductory commentary we present Hartman's classification verbatim.

"In all, we may classify value theories as follows:

I. *Non-Cognitivists.* Here we have
   A. *Empiricists,* who hold that only the empirical is knowable, deny the empirical—or any other—nature of value, and hence its knowability.
   B. *Formalists,* who hold that the value experience appears essentially in value judgments but deny that the logic of these judgments is capable of rendering an adequate account of the experience.

II. *Semi-Non-Cognitivists.* Here we have
   A. *Empiricists* (*Emotivists*), who believe that the descriptive (factual) aspect of value judgments is, but the non-descriptive emotional aspect is not, or not in the same way, logically analyzable.
   B. *Formalists,* who believe that value situations appear significantly in value judgments and that there is a logic of such judgments, but that this logic is *sui generis* and depends on the context of each situation.

III. *Cognitivists,* who are
   A. *Naturalists,* believing that value is a phenomenon observable and analyzable like any natural or social phenomenon. They are in turn divided into

RÉSUMÉ ON VALUES

1. *Empiricists,* who attempt to find value in the subject matter of the empirical—natural and social—sciences, and
2. *Formalists,* who propose to find value through the method rather than the content of these sciences.

B. *Non-Naturalists,* for whom value is a phenomenon *sui generis.* They are divided again into

1. *Empiricists,* who find value in
   a. ontological experience, as an aspect of Being
   b. phenomenological experience, in a realm *sui generis,*
2. *Formalists,* who believe that there is a logic applicable to value phenomena analogous to the one applicable to natural phenomena."

Hartman sees in the ordering of value theories on a scale of increasing rationality, given above, the concomitant and the expression of an evolutionary gradation of value understanding, ranging all the way from the mistaking of value for animal noise to a recognition of its place in reason. Since we shall not have occasion to return to this aspect of our problem, it seems well to pause briefly and sketch the positions at the extremes of this rational classification.

Under I-A are listed the positivists among whom Ayer[12] is perhaps the least inhibited. For him, value judgments are not propositions at all, neither synthetic nor analytic; they are mere ejaculations without reference and hence literal

[12] A. J. Ayer, *Language, Truth and Logic,* 2nd ed., London: V. Gollancz Ltd., 1946.

nonsense. Hence they can not possess normative qualities. Still they can be analyzed according to their purposes in language, though not their meaning. "Stealing is wrong" may thus be a command (Thou shalt not steal), the expression of a desire (I wish you would not steal) or simple disapproval (I don't like people who steal). Considerations of this sort are now common among the analysts of England and the United States who ignore primary experience in favor of language. They tend to forget that literal nonsense can at the same time be non-literal sense. The present view is classified under III–A,2.

At the other end of the scale, III–B,2 among the "cognitivists who are non-natural formalists" we find men like v. Rintelen, to some extent Moore, Tillich, and Hartman himself. The latter defines a good thing as one which "fulfills the definition of its concept" and bases upon this idea a "logic of axiological terms, relations, propositions and truth values" which leads in the end to a hierarchy of values. No doubt these results are impressive and have many of the merits of a formal system. Yet they seem to lack the rules of correspondence which join the system to the unenduring and lowly singularities of human action. Embryonically they are present, perhaps, for if we know the concept of *man* the meaning of good action presumably follows. Or does this argument turn circular? Could it be, for instance, that we should have to know the concept of good action before we can define a good man?

RÉSUMÉ ON VALUES

The preceding survey dealt with value theories in general. A systematic study of theories of ethics, conducted also in the conviction that ethics is an application of a prior and more embracive theory of values will be found in Werkmeister's important treatise.[13]

Another excellent general survey of traditional *ethical* theories which presents each type, together with its representatives, in neat categorization, then appraises its merits and offers criticism, is T. E. Hill's *Contemporary Ethical Theories*.[14] His classification is based on a different principle than is Hartman's, and we offer it here as complementary to Hartman's without further comment.

I. *Ethical Skepticism*
   A. *Logical positivism*
   Carnap, Russell, Ayer, Wittgenstein, Ogden and Richards, H. Feigl, Stephenson (moderate, not complete skeptic), Pap, M. Schlick, T. V. Smith, C. M. Perry are skeptics, but not logical positivists.
   B. *Psychological skepticism*
   Behaviorists like J. B. Watson and psychoanalysts like Freud.
   C. *Sociological skepticism*
   Pareto, W. G. Sumner, Karl Mannheim.
II. *Approbative Theories.* Approving agency is:
   A. *Moral sense (or sentiment)*
   Edward Westermarck, A. K. Rogers, F. C. Sharp,

---

[13] W. H. Werkmeister, *Theories of Ethics,* Lincoln, Neb.: Johnson Publishing Co., 1961.
[14] T. E. Hill, *Contemporary Ethical Theories,* New York: Macmillan, 1957.

A. Sutherland, William MacDougall, A. Shand.
  B. *Society*
    E. Durkheim, Levy-Bruhl.
  C. *God*
    Barth, E. Brunner, R. Niebuhr.
III. *Process Theories* (no unique standards)
  A. *Evolutionary,* involving both cosmic and biological (organic) evolution.
    Cosmic: O. Stapledon, W. H. Sheldon, F. E. J. Woodbridge, F. S. C. Northrop.
    Organic: Peter Kropotkin, C. H. Waddington, Julian Huxley, Haldane, Schweitzer.
  B. *Marxist*
    Marx, Engels, Kautsky, E. B. Bax.
  C. *Pragmatist*
    Dewey, James Tufts, G. H. Mead, M. C. Otto, H. Kallen.
  D. *Humanist*
    W. Fite, C. B. Garnett, Israel Levine, I. Babbitt.
IV. *Psychological Value Theories*
  A. *Hedonistic*
    Durant Drake, R. M. Blake, J. Mackaye, Schlick, W. T. Stace.
  B. *Affective*
    G. Santayana, D. W. Prall, J. R. Reid, C. I. Lewis, J. B. Pratt.
  C. *Interest*
    R. B. Percy, D. H. Parker, F. R. Tennant.
V. *Metaphysical Theories*
  A. *Natural Law theories*
    Maritain, Gilson, M. J. Adler.

B. *Self-realization theories*

J. H. Muirhead, H. J. Paton, J. Seth, J. S. Mackenzie, G. C. Field, W. K. Wright, W. G. Everett, W. E. Hocking, Mary W. Calkins, R. C. Cabot.

C. *Idealistic theories*

W. M. Urban, A. C. Garnett, W. D. Ross, W. R. Sorley, A. E. Taylor, E. S. Brightman, Hugo Münsterberg, N. Berdaev.

VI. *Intuitive Theories.* Values are intuited as

A. *Real*

G. E. Moore, N. Hartmann, J. Laird, A. C. Ewing.

B. *Non-real* (deontological) but nevertheless obligatory!

H. A. Pritchard, E. F. Carritt, C. D. Broad, W. D. Ross, H. Bergson.

The present account of theories of value has been far from exhaustive, conveying little more than a bird's eye view of a highly turbulent ocean of controversy. Perhaps enough has been said, however, to motivate the question whether the value concept with its welter of incongruities can alone form the basis or indeed the starting point of a satisfactory analysis of ethics.

IV

The Working Methodology of Ethics

ABSTRACT

There is a complete parallelism between the formal structure of science and that of ethics. To the axioms of science there correspond the commands of moral systems; to the process of explication, casuistry; to the P-domain, collective human actions. Choice and commitment are required in two places, as in science: first in the selection of commands, then again in the adoption of principles of validation. Confusion between the commands of ethics on the one hand, and its principles of validation (happiness, survival, self-fulfillment) on the other, two classes of elements logically independent, has retarded the clear understanding of ethical problems.

The concerns of ethics stretch from initial commands to the area of human actions. The commands engender specific precepts by casuistic explication; the precepts are lived by a group of men, producing certain actions. In these actions, certain qualities often called primary values appear. Some of these are chosen as validating principles (e.g. Nietzsche's power, happiness, love, self-fulfillment, the peace that passeth understanding, or the ineffable nirvana). If the initial specific commands produce the primary values chosen, the entire living system is a proper, a valid, ethical system, and whatever properties the commands require human actions to possess are called values. The overarching relation between imperatives and their validation creates the normative quality of values.

Statement of Aims

Temporarily this essay will now move away from values and return to them after an important excursion. A focus on values which leaves other matters at the indistinct periphery of one's view will not reveal the crucial elements of ethics in correct perspective. It limits and distorts understanding of ethics as the fact-bound view distorts science. Values do not contain normative qualities as inborn traits; these arise through dynamic transactions and they grow through the interplay of values with something else. Hence the first object of my criticism is an attitude, rather prevalent among value theorists, which may be called axiological isolationism.

In science we found certain normative features originating in the fruitful engagement of various parts which function in the theory of knowledge. Special attention was drawn to the components called P-experiences (facts), rules of correspondence, and the concepts which form theories which, in turn, derive from first theoretical principles; to these principles the names postulates or axioms are usually applied. In declaring a theory "right," reference is made not to a single feature of the epistemological picture but to its entirety; hence the normative in science binds a theory to the full milieu, points to judicious postulation at one edge of our picture and to empirical verification at the other. The normative arises from the integral character of the epistemological process; it dies in the absence of an overarching relation between postulation and ultimate verification.

Now I hold that ethics displays similar formal traits, that one must look for analogs of postulates, epistemic correspondences, P-plane and certain other details in ethics. Somewhere within these is the abode of values. Another parallel will also be drawn with science. That discipline is an evolving, self-corrective enterprise in which what is correct today may not be correct tomorrow. To be sure, changes occur quite slowly, especially near the plane of postulation,

THE WORKING METHODOLOGY OF ETHICS

but they do occur. One should expect, perhaps even welcome, such a dynamism in ethics.[1] Finally, let us note that there is no single all-embracive set of postulates for science, nor is there one unique method of verification. Hence, one should expect similar complexities at the postulational stage in ethics, too. The attainment of one final scientific theory, applicable to all cognitive experience, is a matter of hope and striving which animates researches and puts a gleam into the eyes of many scientists. But science is possible before this final goal is reached, possible indeed if it is never reached. And so it may well be with an all-embracive and ultimate system of ethics. Ends can be useful if they are never gained; stars guide the sailor without being attainable; Columbus never achieved what he set out to do, i.e., facilitate trade with China, but he discovered America while seeking that end.

[1] To alleviate the shock of this assertion I note here that it is not really very radical or novel. Even those who see in Christ's admonition, "love thy neighbor," a so-called absolute and unchangeable precept will admit that the understanding of that imperative, and, therefore, its effective meaning changes in time. Christ himself, when asked "who is my neighbor," told the parable of the good samaritan, answering effectively, anybody in need. The idea of neighbor has undergone many interpretations; even in our day there are those who confine it to friends and exclude foes ("Praise the Lord and pass the ammunition") and those who wish to include all men.

Equally wide variations are to be noted in the meaning of love, which can mean actual fondness, i.e., positive affection or merely a benevolent disposition. To this point we shall return in Chapter VII.

Every living doctrine of ethics is, and should be, subject to enlarged understanding in conformity with new evidence and with changes in the human situation.

To clarify further the position here developed, I plead guilty at once to the charge of using science as a model for ethics, albeit as a formal, methodological model. For this I would cite an attenuating circumstance: I share the realization which grows alarmingly in many minds, that ethics, moored in a stagnant but pleasant pool of values, amid the enchanting phosphorescence of decaying ideals, is practically ineffectual, especially in the international sphere; whereas science has acquired a measure of success and an efficiency which frightens us. In using science as a model, however, I do not take the facts, nor even the specific features of scientific methodology and transplant them into ethics. Things are not as simple as that. The task ahead is to *find* their functional counterparts.

Above all, lest this point be misunderstood, the scientific *is* will not be inflated into an *ought*. While we have discovered quality in science (its being "right" or correct) we have also recognized it as logically and operationally different from the normative in ethics. This difference, it will appear, is so radical that no combinations of scientific propositions can ever yield an ought. Here we part company with the language analyst who could not concede this point. For language shows no cause why this difference must be taken as a radical one. Out of declarative sentences one *can* build an imperative without loss of meaning. Phrases like "stealing is improper" are almost exactly equivalent linguistically to "do not steal." Yet there remains a healthy sense of difference between these phrases, a sense borne out by science as we shall see. The apparent similarity again merely indicates that language is an imperfect image of experience.

Science is indiscriminate with respect to the ethical ought, but it may aid or hinder its realization. It provides *facilities* for all kinds of action. It enables man to preserve life and to kill, but it never tells whether it is morally[2] right or wrong to preserve life or to kill. It may even show killing to be inhuman in the sense of being abhorred by most men. This still does not bridge the gap between being inhuman and being ethically wrong. It may show that preservation of the race is the goal of evolution (which one may doubt); yet to hold that human actions must (ought to!) conform to the end result of evolution is a separate (and, I think, erroneous) belief.

Analogy, methodological similarity between science and

[2] In this book we make no distinction between the meaning of "moral" and the meaning of "ethical," nor between morality and ethics. Other writers insist on careful discrimination. Professor Errol Harris, to whom I owe this caveat (private communication), has this to say. "Morality consists of the 'mores' of a people. The moral rules they recognize, the imperatives they impose and obey, the practical standards they apply and the conventions they accept. One may, of course, use the term generally to apply to the practical regulation of life according to moral rules of our people or culture, or may use it to refer to a special system of morals in a particular community. Ethics, on the other hand, is the (philosophical) theory of morals, the object of which is the elucidation of moral criteria, of the source of moral obligation, the critical assessment of standards, and the understanding of the bearing of the notion of goodness on the obligatoriness of rules . . . and so forth."

To cover this distinction we have preferred to speak of factual and normative values, of *est* and *esto* laws, of is and ought.

ethics are supportable relations; equivalence and reducibility are not.

In this chapter we introduce and examine the ethical counterparts of axioms, protocol facts and rules of correspondence.

"Axioms" of Ethics

We begin with a brief comment on the recent history of axioms.

Wherever they are acknowledged as important factors of thinking, e.g., in logic, mathematics and physics, their status has undergone a profound change during the last two centuries. They entered the scene with authoritarian trappings, as truths proclaiming themselves certain, absolute and therefore indubitable. Their affidavits bore the seal of divine ordination, of some *lumen naturale,* or the light of inner reason; more technically they were known since Kant as synthetic *a priori* judgments. Such authoritarianism has now largely disappeared where it once existed, and we must take note of this before we look for axioms in ethics.

What has happened is most readily illustrated by a famous example from geometry. It is a tired old story, invariably employed in this connection, and I use it here with some hesitation; still it is difficult to find another instance equally appropriate for the lesson I wish to draw. Euclid's geometry contains five axioms and five postulates in its original formulation.[3] It is permissible to gloss over the distinction Euclid

---

[3] For a succinct account, see Lindsay and Margenau, *Foundation of Physics,* New York: Dover Press, 1957.

intended between axioms and postulates and to say that there were ten postulates in all. The last of these, the so-called parallel postulate, asserts: If two straight lines in a plane meet another straight line in the plane so that the sum of the interior angles on the same side of the latter straight line is less than two right angles, the two straight lines will meet on *that* side of the latter straight line.

To the successors of Euclid it seemed that this postulate, because of its rather specific and incidental nature, was not an independent necessary proposition, and they set out to derive it from the other nine. In this they failed, concluding that the statement was indeed a postulate. Strictly, of course, they had merely produced evidence for its independence, not for its necessity or its truth. The more daring mathematicians began to suspect the last named qualities, and in 1832 the Hungarian Bolyai succeeded in showing the formal dubitability of postulate ten. He replaced it by another, quite different in content, denying the postulate in question, and proved that the resulting geometry remained consistent and equally fertile in the production of theorems. Nor was this particular substitution unique; others followed (e.g., Lobatchevski, Riemann) and a whole series of non-Euclidean geometries was produced.

More remarkable than this formal feat was the discovery, begun by Einstein and Minkowski, of the possibility that these geometries may have empirical applications. Very likely, present empirical knowledge about the motions and distributions of galaxies in space is more readily interpretable on the assumption of Riemannian than of Euclidean geometry. All this suggests coercively the failure of the original authoritarian claim concerning postulates: A set of them may seem consistent—although even this can no longer be certified—but they are not for that reason unique or necessary; in particular, their seeming consistence and, to put it properly, their mistaken uniqueness does not of necessity render them applicable to the "world of experience." Absoluteness, apodictic univocality has gone out of them and they have been reduced to what the Latin *postulare* denotes: important premises which we "beg" experience to confirm. In more solemn phraseology they are scientific articles of faith to which a thinker commits himself until contravening facts release him from his commitment. In Chapter I we analyzed the tentative character of axioms from a logical point of view; to affirm them involved the so-called "fallacy of the consequent"; but, paradoxically, the progressive deepening of scientific understanding and the flexibility of science are benefits for which we are indebted to the new modesty in the claim of postulates.

Decrease of assurance from certainty to provisional tenure is evident in another movement which dominates present philosophy of science. Since Kant, the synthetic *a priori* has fought a retreating action against the onslaught of science. Synthetic *a priori* statements are supposed to express universal and necessary truths. Time and space are infinite, $ab = ba$ in algebra, every event has a cause, all matter is extended, matter is conserved—these and other general statements were once held incontrovertible, true and neecssary. One after the other, they have been questioned and finally contradicted by the science of our day, and it seems no longer reasonable to affirm such dicta in the Kantian way. To be a Neo-Kantian generally means to see this point but to be unwilling to admit its full significance. What remains of the *a priori* is the concept "logically prior"; this depends on the structure of a given logical system and is therefore relative. A proposition can be *a priori* within a specified context of statements and facts, but not without any qualifications.

So much for the axioms of science. In ethics, the desire for absolute assurance at the point of the axioms, whatever they may be, is still very strong. The waves of the scientific revolution have not yet reached the shores of the island of ethics except for the off-shore stations manned by positivists, which have been inundated. Moreover, the isolationism of the islanders has erected breakwaters to keep them off. The reverential moralist refuses to accept principles unless they are indubitable, sacrosanct, and eternal; on the opposite side the immoral person regards himself unbound by laws of ethics which cannot be proved. The effectiveness of ethics seems to depend upon the existence of absolute and universal truths.[4] Consider against such attitudes the patent fact that

[4] For an elaborate examination of the effects of the removal of the absolutes once provided by "high religion," absolutes which in his phraseology are being dissolved by the "acids of modernity," see Walter Lippmann, *Preface to Morals*, New York: Macmillan, 1952.

science, odious as it may be to many, has flourished immensely under the regime of tentative postulates!

When complete assurance is sought it is easiest to rely on divine revelation. Doubtless this is one of the reasons why most historical systems of ethics are closely coupled with religion. Another lies in man's striving for justice, his feeling that justice must rightfully prevail even in the face of irrational incidence of evil, unmerited happiness and wanting retribution for sin in an individual life. This feeling cries out for an unseen judge, for fore- and after-lives which fulfill the demands of award and punishment. The same feeling can, of course, lead to Promethean, anti-religious defiance of God. A cool and objective appraisal, however, can hardly miss the patent conclusion that an intimate connection between ethics and dogmatized religion bespeaks an immature, unformalized understanding of ethics, even if the connection adds power and workability.[5]

Historically, every one of man's deep cares was once clothed in religious myth; science was no exception. Then, as inquiry and thought develop, as special problems appear and are found capable of solution, the religious element weakens and sometimes disappears completely. It is my belief that it returns in a finer, more articulate and responsible

---

[5] The popular confusion is exemplified by the story of a minister who interviewed a parishioner and received the complaint that the parishioner had trouble with his belief in God. The minister, after listening to his story, finally asked: "Don't you really have trouble with the Ten Commandments?" People are prone to justify their lack of morality by pretending a rejection of religious beliefs.

THE WORKING METHODOLOGY OF ETHICS

form when a higher degree of perfection is reached in any field of inquiry, partly in response to those miraculous features of the universe and of life that make man's understanding possible. Ethics, however, has not yet entered its second stage in Western Culture, the stage in which it frees itself from outright religious domination.

Perhaps this judgment is a little severe; for certainly the coupling of ethics with religion in the Western World has also a visible cause in history. The principal figures of Judaism and of Christianity performed double duty in laying the groundwork not only of religions but also of ethical codes. Hence both grew side by side, grew into each other and remained connected in the Western mind. Yet a separation is possible, for there are many Christians who reject religious claims and regard Christ as a successful moral teacher. Buddhism, Taoism, Confucianism are basically ethical systems in which the emphasis on religious doctrine is weak, and it is interesting to note that intellectual Buddhists, when they are converted to Christianity, give as the most attractive quality of their new beliefs the profound moral insights of Christ.[6]

Ethics *can* stand on its own feet, and it should be weaned from religion to gain strength. It can do this without loss of love or disrespect for its mother. As to the arguments advanced in this book, they leave the historical genesis of our

---

[6] See, for instance, Lin Yu Tang in *Christian Perspectives in Contemporary Culture,* ed. Frank S. Baker, New York: Twayne Publishers, Inc., 1962.

ethical convictions undetermined and open to many causes, treating it as we do the injection of great ideas into the stream of science, where we speak of genius and inspiration and also of the creative force of circumstances. At any rate, I, myself, think it unreasonable to confine divine inspiration to ethics and religion. Either there is none anywhere, or else I like to see it acknowledged in all areas of human creativity.

We now come to grips with the central question: What are the "axioms" of ethics, the starting points of all ethical procedures? Their nature should at once be evident from the task they are meant to perform. In science axioms serve the purpose of describing and explaining factual experience, of portraying what was, is, and will be. Their mood must, therefore, be the *indicative*. Ethics means to guide motives, wills, and actions; hence its axioms naturally speak the *imperative* voice. We conclude, therefore, that the axioms of ethics are commands, and this conclusion finds universal confirmation in the prevalence of commands among the basic tenets of successful ethical systems. "Thou shalt" reverberates through history as the greatest moral force; values pale before it, sicken, and remain alive in the nursing homes run by moral philosophers.

Some value theorists claim that standard logic[7] is not applicable to values. This is evidently not true for imperatives,

---

[7] My tendency here is to include modal and deontic logics among standard forms. But I am not even certain that ordinary logic can not handle imperatives.

which do allow standard logical transformations. They can be contradicted; they may or may not have an excluded middle; imperatives can be categorical, hypothetical or disjunctive, and the categorical ones can differ in accordance with the old rules in regard to quality (they can be affirmative or negative) and quantity (they can be universal or particular). Hence it does not appear that a new calculus is needed. Inferences can be drawn in the good old ways. From the universal command, do not kill, innumerable specific commands follow if the meaning of the verb to kill is understood. A detailed examination of specific instances often aids the understanding of the universal, just as is true for indicative or declarative propositions. In a certain sense the imperative is even simpler than the indicative, for it lacks the existential referent which imparts grave problematic features to every indicative sentence.

If, as some proponents of axiology (e.g., R. Hartman, *loc. cit.*) maintain, a new kind of logic were needed to deal with normative injunctions, that need should have shown up in jurisprudence. As a matter of fact, judges continue to reason in the forms of Aristotelian logic. Clearly, there seems to be no call in this area for a modified logical system. This might have been expected on very elementary grounds: The distinction between indicative and imperative is a linguistic one, and it does not go very deep into our modes of thinking. Even in the Indo-European languages every imperative can be translated into an indicative sentence by the use of the phrase "It is proper that . . ." Like so many features of our thinking, this distinction is strongly conditioned, if indeed not created, by the peculiarities of our language.

We notice, too, that the use of the imperative gives us a preliminary *formal* hold upon the ought, a convenience not provided by the indicative which is appropriate to values. What the ought lacks for completion is a person's acknowledgment of the imperative, his commitment to the command, and this is the analog of the (tentative) acceptance of postulates elsewhere. How this acknowledgment is gained is another question to which we turn in the later parts of this chapter. Suffice it here to say that commands can be accepted for a great variety of reasons, among them love or respect for the giver of the command, simplicity or plausibility of its directive and the *effects* it produces among people who follow it.

In contrast with *much* contemporary literature on ethics, *the record of truly creative ethical works is notable for its use of imperatives and for its being specific.* I believe this to be an honest judgment, admitting that it is difficult to substantiate because of the confusing texture of extraneous material, cosmological, religious, and philosophic, which is so often woven into the fabric of moral systems in order to provide motivation and authority for commands. It may also be that my own personal bias had added weight to these two features, use of imperatives and specificity, during my search. Abstract value theory was introduced to Western thinking by a rationally powerful and artistically gifted, but ethically rather mediocre culture called ancient Greece, whose legal system was primitive and unorganized. Here arose the controversy about the meaning of the good, the doctrine that virtue is knowledge, that the good is the ultimate ground of being, or the pure activity of universal abstract thought; to some extent this argumentation continued in Rome, where the Stoics saw the good in wisdom, the Epicureans in pleasure.

It is fair to say, I think, that the really massive contributions to living ethics have come from elsewhere, and it is in the instances cited below that specific imperatives prevail. The decalogue of Moses need not be expounded here. A similar code, with very similar mythical trappings, was given to Babylon by Hammurabi about one thousand years later. Zoroaster, after another millennium, presented to the Persians a religion whose core is ethics. There is a dualism between heaven and hell, or good and evil, each personified by a deity. But in the ghatas, which are taken to be the verbal account of the prophet's teachings, Zoroaster speaks of divine *commands*. The abstract good, symbolized by Mazda, is exemplified by seven particular virtues, and these are personified as genii watching over specific precepts; among them are truth, obedience, reverence for the divine, and good sense. Buddha, in his doctrine of the eightfold path, spells out what man must do to attain nirvana, and he gives little in the way of rational justification.

The Tao is full of admonitions in which imperatives are coupled with the specific ends they will achieve: Be humble, and you will remain yourself. Be flexible, bend and you will remain straight. Be ever receptive and you will be satisfied. Become tired and you will be renewed . . . In Confucius, whose ethics is unusual for his age because it lacks a grounding in religion, we find the first statement of the Golden Rule (in negative form, to be sure): Do not do to others what you do not wish others to do to you. And this is followed by an anthology of ethical precepts covering a range of behavior from the dignity of a superior man to the proper style of speaking. The Golden Rule is repeated five hundred years later by Jesus of Nazareth (Luke VI): And as you wish that men shall do to you, so do you unto them. But all the psychological vagueness which remains in this profound maxim is dispelled in the Sermon on the Mount, which is full of specific imperatives. Some of them are couched in the beautiful "blessed are they who," which is a gentle command, others speak straightforwardly in the form "I say unto you," and couple the injunction with promises of retribution. In Mark, Jesus gives the supreme ethical command, "Love thy neighbor as thyself," which may, in a certain sense, be regarded as a Christian extension of Confucianism.[8] Only the gospel of John treads lightly on imperatives; here all good actions are demanded for the personal love of Christ, and ethics recedes behind a wealth of religious feeling.

[8] For a more careful analysis of the Love doctrine, see Chapter VII.

THE WORKING METHODOLOGY OF ETHICS

This brief survey of commands in history was here intended to show their prevalence, their similarity, and their force. Some readers will object at this point that ethics, when based on arbitrary imperatives, degenerates to law. Let us, therefore, consider this allegation. As to the *arbitrary* nature of the codes, comment will here be withheld; later, however, it will be shown that the commands are almost never the whims of individuals or the autocratic decrees of rulers; they are intuitive gleanings from the field of human experience which is rich in confirmation of ethical principles, now and then culminating in celebrated pronouncements by men of deep insight. But they are, nevertheless, commands and hence like laws of uncertain force.

What, then, is the difference between ethics and practical jurisprudence? I hold that there is none in principle, that the two are related like science and engineering. One devotes itself to fundamental and general problems, the other to specific situations occurring in daily life; one is largely left to the care of the individual, the other is a public concern because the immediate safety of society rests upon it. Murder is a moral issue, but it must also be prevented legally for the same reason that we can not tolerate the collapse of a bridge, that we need traffic laws and paved streets. The same *basic* urge which produces the conscience in the heart of man produces law in his society. Indeed the philosophy of law involves the same problems as does moral philosophy and often, because of practical urgency, the legal discussion is clearer and more to the point.

Ethics and law *appear* different for three reasons. In the first place, laws are visibly enacted and their origin is usually known, while ethical maxims often, though as we have seen not always, grow unseen. Secondly, legal statutes make provision for the punishment of violators, most ethical injunctions do not—only in some religious systems are law and ethics merged by the concept of retribution in after life. According to the view here presented ethics, too, looks for rewards of a subtle kind; the command is never blind, as we shall see, and the similarity is not wholly absent on this score. The reason for the greater emphasis on individual punishment in violations of legal statutes, however, lies in necessity and sometimes in a sense of revenge; to exist at all organized society is practically forced to punish acts of tampering with its functions. This point of view becomes increasingly dominant in the development of Soviet law. Thirdly, legal action covers cases in which it is felt that there is no violation of ethical principles at all. Being tardy in a tax payment or going through a traffic light at 2:00 A.M. when the streets are empty, such are infractions of legal statutes which do not ordinarily count as violations of moral law. But perhaps we indulge here in sophistry, for if honesty and justice are moral obligations, so is the payment of taxes on time; and if the preservation of life is an ethical imperative, then risking it needlessly by running through a traffic light is also unethical.

Laws, statutes, regulations seem somehow too particularized to be of ethical importance. But this semblance of moral insignificance arises also when the consequences of an acknowledged ethical imperative are spun out too finely. The minute implications of the demand for complete honesty or veracity can under imaginable circumstances prove embarrassing and lose all moral aptness, and on the other hand a friendly gesture or a greeting short of complete sincerity may take on moral value. There are large areas of human behavior where ethics and law become artifacts, and this is as it should be.

In jurisprudence it is recognized that too much legislation is an evil; overdetermination of action by ethical codes is equally vexatious and pointless. Since the purpose of this book is to help reveal what ethics and science have in common, let us recall here that it would also be foolish to expect or promote the control of all affairs by science. As a matter of fact, no scientist has such expectations. He knows that there are many matters that lie beyond his scientific ken, and he hopes that this will always be the case; for if it ceased to be true, there would be no more science. Fortunately, the forms of experience are richer than the principles of reason. And in the same vein we expect the complexity of human behavior to transcend the governance by ethical imperatives.

We have identified the starting point of ethics, the counterpart of axioms in science, with commands, with imperatives, and considered what they mean and do. That they can not be proved in *a priori* fashion is no longer a mystery in view of the recent fate of postulates in science. But we have not examined their origin except by way of examples, and we have said nothing about the way in which they attain validity, nor what it means for them to be valid. We now take another step toward an understanding of these pivotal problems.

To loosen the discourse and conform to custom, I shall henceforth often speak of these imperatives as *codes* or *norms*. They are, of course, not values; they are the generators of values. And it will be of paramount importance to keep them clearly separated from another component which is present in every functioning ethical system, namely from the principles of validation.

EXPLICATION OF AXIOMS

In science, the content of axioms is made manifest by deductive logical or mathematical procedures, by stepwise explication. From the postulate of universal gravitation, coupled with Newton's laws of motion, one obtains differential equations which can be solved in a way that leads to Kepler's three laws of planetary motion. Much of the information needed to put a satellite into orbit is spun out of them by careful mathematical procedures. Explication, the exposure to view of everything an axiom implies, is logically nothing more than the establishment of tautologies. The flavor of that word condemns explication to insignificance if not impropriety in the mind of the casual bystander, yet every scientist knows how important and how difficult it is to know all that is implied by postulates. Except for the statement of a few postulates a book on theoretical physics is nothing more than a compendium of analytical procedures showing forth what the axioms contain.

In ethics, explication of norms is *casuistry,* and this word also carries the stigma of uselessness or impropriety. Mainly for this reason moral philosophy has become practically impotent; a course of study in it has no bearing on a student's ethical attitudes; it is purely theoretical, enlightening but not morally educating. Instead of regretting this, teachers of ethics sometimes try to make it seem normal, and to give some plausibility to this attitude they call attention to courses in history which do not change history, or to courses in poetry which do not pretend to enable the students to write poetry. Only science, engineering and the fine arts include the specific and detailed application of principles (i.e., "casuistry") in their teaching. I hold that ethics should be like these disciplines in point of practical efficiency. And casuistry, though distasteful to most experts in ethics, must be included as an important branch of the subject's methodology.

In a speech before the Jewish Theological Seminary of America (reported in the *New York Times* on November 11, 1962) Chief Justice Earl Warren gave forceful expression to the need of modern explication of ancient Biblical norms.

He called for "counselors in ethics" to help Americans apply moral principles to the complex problems of modern life.

"I do not regard the word of Scripture as a dead letter, addressed only to the generations who heard it from the mouths of the prophets," Warren said. "I regard the Scriptures as a living tradition, as applicable in our time as in any other.

"But in a changing world, this world needs new interpreters, adventuresome spirits able to make it effective in our lives."

Casuistry has one purely logical component which likewise tends to be neglected; writers on ethics seem never to worry about logical consistency of their systems' imperatives. Consider, for example, the Ten Commandments (which we enumerate here in the old-fashioned order in which the fifth says "Thou shalt not kill," the seventh "Thou shalt not steal"). Certainly the first three are not ethical at all; they are religious and ceremonial in their aims. The ninth is logically a special case of the tenth.[9]

Other historical codes, notably the Tao and the Analects, are much more proliferous and present a far greater logical redundancy and at times confusion. Whether there are conflicts between the various utterances of Lao Tse, for example, seems to be of no interest but consistency is always

---

[9] When in these pages we speak of the decalogue, reference is always intended to the *ethical* content of the Ten Commandments.

tacitly assumed. Systematic study of ethics should clear this up.

Aside from the logical role of casuistry there are the more familiar ones of clarification and persuasion, as exemplified in catechisms.

There are good historical reasons to explain why casuistry has fallen into disrepute in Western societies. As the art of bringing general moral principles, especially moral codes, to bear on particular actions it was once a flourishing discipline. The decalogue, for instance, is followed historically by the development of detailed Jewish laws and by the writings of the Rabbinical schools which are highly casuistic. Here, interpretations of the law reach into absurd details of daily living such as forbidding the killing of insects on the sabbath day.

A noteworthy exception to the use of casuistry again is ancient Greece,[10] about which Butcher[11] has this to say. "No system or doctrine and observance, no manuals containing authoritative rules of morality, were ever transmitted in documentary form. In conduct they (the Greeks) shrank from formulae. Unvarying rules petrified action; the need of flexibility, of perpetual adjustment, was strongly felt." Their reasoning, it is true, was not always above attention to minutiae. But instead of taking the form of casuistry their arguments

---

[10] M. R. Thamin, "Un Problème moral dans l'antiquité," *Extraits des Moralistes*, Paris: Hachette et Cie., 1897.

[11] S. H. Butcher, *Some Aspects of the Greek Genius*, London: Macmillan and Co., 1891.

concerned what the Greeks regarded as points of principle, e.g., whether a good man is good when asleep (Aristotle).

When stoicism took hold in Rome, where ethics had always tended to take on the more rigid form of legal imperatives, casuistry developed rapidly and became the concrete side of moral philosophizing. Cicero and Seneca were masters of this art. The latter emphatically made the point I have here attempted to convey: that the practice of spelling out the meaning of moral commands in particular situations gives an interest in morality to those who "have no love for abstractions," and insistence on the concrete prevents those who have such inclinations from losing themselves in the clouds. Should a merchant reveal defects of his goods to a prospective buyer? May a lawyer defend a client if he knows him to be guilty? Am I to lie to save someone's life? These are sample questions to which the Roman stoics turned their minds.

Christianity, in spite of its accent on love and charity, on good disposition rather than good deeds, developed casuistry into a formidable discipline in connection with the sacrament of confession. Priests were required to impose penances; to facilitate this task they developed *"libri poenitentiales"* along with graded registers of sins. It was against these technicalities of the moral industry that the Reformation raised its sights, and in the ensuing struggle casuistry lost its prominence and its reputation. That art was replaced by reliance on the individual conscience, the moral guardian within man. The Counter Reformation attempted to reintroduce casuistry in a form much weakened by the theory of "probabilism" (enjoining priests to be very lenient in their appraisal of sins and easing the attainment of absolution) and succeeded in a measure rarely recognized by the Protestant church. In protestantism casuistry arose only in movements now regarded as atypical, for example in the puritanism of our own early settlers for whom a *working* system of ethics was a practical necessity.

No new logic of values was used in these instances of moral explication; the connection between axiomatic commands and specific precepts for action rested on the old Aristotelian two-valued calculus, which served its purpose adequately. This purpose is twofold: first, to make known and concrete the meaning of the primary code; second, to appraise it for its sufficiency. The second is a methodological requirement whose importance will be clear later: like the postulates of science ethical commands must be consistent and in a certain sense complete; they must neither overdetermine nor underdetermine observable behavior. More specifically, they must not allow actions which common understanding deems unethical, nor disallow clearly moral acts. Only by an application of casuistry, properly understood as moral explication, can these purposes be achieved; hence, we need this activity, and I think we need more of it than is currently practiced.

The direct form of ethical explication considered thus far,

which we thoughtlessly tend to reject, is supplemented in all cultures by highly esteemed indirect methods of casuistry, not ordinarily recognized by this name. We refer here to heroic sagas, the great epics that have molded ethical conduct since the dawn of history, and which were in essence anecdotal interpretations of moral principles. It is true, we look down upon tales that end in obvious moral lessons or carry too thick a spread of preachment on too thin a slice of literary workmanship. But in spite of our insistence on unembellished realism, the best works of literature to my mind are still those that grip the reader's moral fiber.

The Protocol Plane of Ethical Experience

The protocol experiences for science are data, observations, those matters which we loosely call facts. The literal meaning of *"factum"* (that which is made) is precisely the opposite of what the word fact signifies in this context: facts are supposedly *not* made by man, they are independent of his knowledge, his reasoning, and his interpretation; the word *data,* (that which is given) or even *habita,* (that which is had) characterizes their status more correctly. Facts are beyond our control except insofar as we can consciously attend to or ignore them. Even when we contrive their occurrence, as we do in experimenting, measuring, designing, there remains a sense in which the facts engendered are not made by us, are not at our mercy.

As we have pointed out in Chapter I, the data as subjective experiences are accepted without doubt in every science. One may question their meaning, their relevance, their reference to actual objects, the veridicality of reports concerning them, but not their being experienced in first-person consciousness. As spontaneous, indubitable, coercive phenomena they form the final criteria for the correctness of all scientific theories and speculations. Hence our term, protocol (P-) experiences. To be sure, the P-experiences for different sciences are not the same. Feelings belong to the P-plane of psychology, rarely to that of physics. Observed happenings always form the P-experience of natural science. History is in a peculiar situation insofar as it concerns itself with the past which is no longer remembered; historical events are not to be found on its P-plane, they are the constructs (cf. Chapter I) of history which are inferred from inscriptions, manuscripts, eyewitness reports, etc. These latter form the P-experiences of history. Are there also protocol experiences for ethics?

These, we affirm, are observed or historically established human actions. For reasons to be discussed later, individual actions will be excluded from consideration unless they occur in situations unique enough to be ethically relevant. If this restriction to *collective* action seems unwanted and troublesome we can admit individual acts to the domain of P-experiences of ethics but treat them in the same manner in which the scientist treats single, unrepeated observations.

The reason for the choice of observed human actions to

play this role of P-experiences is twofold. First, they are peculiarly well suited for this crucial role as will appear when the outline of the present methodology is completed. Secondly, it seems that the unindoctrinated, every-day practitioner of ethics does in fact look upon the milieu of actions around him, especially the actions provoked by his own behavior, as important and often ultimate criteria of the success of his ethical precepts.

A radical distinction arises from the circumstance that the protocol material of science comprises facts, while that of ethics is composed of actions. Actions, as mentioned, often involve freedom of choice, an element which is foreign to the field of facts. Hence the very nature of the protocol domain of ethics requires a kind of ought which would be meaningless for science. Ethical actions *require* it; they are empty in a way facts are not when the ought is missing.

Within collective human behavior one may discern notable states of affairs of qualities (natural qualities, to employ what seems to be widespread terminology). These include social chaos, civil strife, to name a few that are unpleasant; cultural longevity, happiness, freedom, health, wealth, self-fulfillment, altruistic love, education, refinement, self-denial, austerity, peace, saintliness, on the pleasant side. Strictly speaking none of these is directly observed, and if one wishes to push the parallelism between science and ethics to the very limit one may say: these qualities are related to human behavior by epistemic correspondences, just as temperature, electric charge, chemical valence, etc., are related to immediate observations. If ethics were as highly developed as natural science, its philosophers would doubtless wish to analyze the correspondences with the care scientists have employed in pursuing this task. They might, for example, find counterparts of operational definitions for happiness, self-fulfillment, and the rest. These qualities would then become measurable, and this would enhance the precision of their meaning. Whether this proves feasible is not open for conjecture, the question needs to be settled by empirical means, and modern sociology is clearly moving in that direction. One must applaud these efforts while realizing at the same time that ethics, conceived from the present point of view, remains possible as a qualitative discipline even if they fail.

The qualities in question are traditionally called values (or disvalues). As we have seen, that word is not reserved for them but is lavished upon everything else that seems desirable, including the attractiveness of material objects and indeed things themselves. Some call these qualities primary values because they seem very basic and give, as it were, secondary value to other things. If this terminology is used it must be remembered that they do not carry normative force as "natural" attributes so long as they are mere qualities of human behavior. Whatever force they acquire results from our further acknowledgment of them as "good" or "right," from a personal commitment to them. So far, they are collective qualities apparent in human behavior and nothing more. We shall see that they play an important role in the validation of ethical codes; this problem is to be discussed in the next section.

The Validation of Norms

One part of the view I am developing is that ethics is an empirical science which needs to be tested in experience. The concepts of science are tested for their correctness; they are verified. The concepts (commands and explicit precepts) of ethics are tested for their adequacy; they are validated. Correctness turns certain constructs of science into *verifacts;* adequacy turns certain constructs of ethics into *values.* Through these transformations they acquire *oughts,* normative qualities which lift them above the statistical and the subjective.

Against the view that ethics is an empirical science stand several weighty objections. One is sentimental, the other rational. The sentimental objection invests moral matters with an aura of high dignity and sanctity, in consequence of which it refuses to have them dragged into the dirt of common experience. It is said that ideals are made of nobler substance than the world; how then can the world ever bear witness to their worth? Innumerable poets have fastened on this theme, but not all. One of the greatest among them, the German Goethe, saw through the shallowness of this argument and turned it about, saying

> "Wär nicht das Auge sonnenhaft,
> wie könnt es je die Sonn' erblicken?"

THE WORKING METHODOLOGY OF ETHICS

If the stuff of the world were not itself inspired by ideals, how could we react to them? Neither form of the argument is cogent, and we may take our choice. I prefer the latter.

The rational objection has considerable force. Affirming that values are appetitive to certain goals like pleasure or happiness, this contention goes on to say that nothing can be gained by empirical confirmation. For it is clear that action designed to achieve the goal of happiness will, if circumstances are properly arranged, attain this goal, and if it fails, the indictment is not directed against the value which motivated the action, but against the factual situation responsible for the failure. Actions seeking values rubber-stamp the already assumed validity of these values when successful, prove nothing when they fail. True empirical validation is impossible.

If values exhausted the concerns of ethics, and if they drew their entire substance from the appetition of goals, then it is true that these goals have served their purpose in defining values and cannot be drawn upon for further discrimination or use. But in the view here presented neither of the premises is true; values are not the sole concern of ethics, and their content is not determined fully by goals. They are suspended, as it were, between the axiomatic commands and the qualities of collective behavior and receive their meaning from both. Our story, instead of beginning with values and going on from there to happiness, health, love, etc., in admittedly pointless fashion, starts with *com-*

mands, uses values as mere stepping stones toward observed behavior. Beginning and starting points are independent, *for it is empirically clear that a given set of commands will not necessarily lead to a specific selection of qualities in observed behavior.* Thus, the decalogue was doubtless conceived by Moses with definite purposes in mind: perhaps the survival of the Jewish people, or their collective happiness, or the satisfaction of Jehovah. But these results could not be guaranteed as automatic consequences of the code; their incidence, while hopefully expected, needed to be demonstrated by living in accordance with the decalogue. Hence compatibility between imperatives and the desired human results is not trivial nor assured and there *is* opportunity for empirical validation.

Aside from the claim that ethics cannot be an empirical science, a challenge I have just tried to meet, one encounters the belief that ethics is completely empirical and utterly wanting in rationale; hence it is thought that validation is impossible because there is nothing to be proved. The subject is thus likened to a descriptive science like geography, where the ascertainment of facts concludes one's tasks. The comparison is accurate if ethics is nothing but a symphony of feelings full of sound and fury, or if it is a meaningless assemblage of linguistic forms. Both of these interpretations are here denied.

When the possibility of validation is granted one must face the question of particulars: What are the qualities of the

P-plane which action in accordance with a postulated code and its logical consequences must engender? Are they survival of society, or wealth, or personal pleasures, or happiness, or self-fulfillment, or indeed self-denial? Clearly, these often conflict, since a course of action achieving one may destroy some other quality. Here, then, we seem to be meeting our Waterloo, for only through arbitrary selection can we set our ethics in motion. The commands are of no help, for they are presumed to be independent (at least in principle) of the happenings on the P-plane, just as the axioms of science are (logically) independent of the facts they are expected to explain. Nor can the factual consequences of the commands be foreseen in *a priori* fashion.

Let us now ask a very simple question which may startle the reader: What is embarrassing or logically offensive about the need for arbitrary selection at the juncture now reached in our ethical epistemology? Is acceptance of *choice* at this point too high a price for avoidance of the self-reference of values? To me it does not seem like an exorbitant price at all but like an added benefit, permitting man to make a wise decision. The locus within the methodology where this second posit is required is likewise of significance: Our first commitment was to an unproved code which stands at the start of all ethical procedures, the second comes at the end, at the point of validation, where it must seize upon some *principle of validation*. Suppose a logical system possessed only one set of basic postulates, as Euclid's geometry did.

Once they are accepted, all theorems derived from them are analytic, tautological, and the entire system is predetermined, rigid, and inflexible. A system of this sort is unwanted both in empirical science and in ethics. To escape analytic rigidity, at least one further element of choice, an independent postulate must be introduced elsewhere. In science that postulate links constructs to observations and fixes those conditions under which agreement between prediction and measurement is said to exist; in ethics it enables validation.

In the last sentence I have used the singular, postulate, as a convenience. Strictly speaking a postulate, like an idea, is never single, for it can always be divided into component postulates or ideas, much as a line segment can be subdivided in innumerable ways. It would be better, perhaps, to speak of a cluster of postulates at the point where science verifies and also at the point where ethics validates.

We now recall the discussion in Chapter I (page 49), which dealt with verification.

In science, this cluster contains first of all the very basic stipulation of what might be called its purpose: science intends to explain, or foresee, factual or datal happenings. It does not want to predict inner feelings or, as we have seen, establish values. Next, it commits itself to explaining or foreseeing factual happenings within a certain tolerance set by arbitrary convention; the act of foreseeing or predicting involves rules of correspondence, and the range of this tolerance we have recognized in Chapter I as the permissible error.

Having made these choices science goes ahead and succeeds. There are certain intrinsic technical considerations, coming in this instance mostly from mathematics (theory of errors), which provide aid in the latter choice, and science happily avails itself of all the help it can get.

In ethics the situation, if formally different at all, is perhaps even simpler. *Its purpose is to motivate actions,* not to explain facts or to predict inner feelings. Secondly, it does so by selecting from the qualities exhibited by collective human behavior a special one, or a set of them (e.g., power over others, personal happiness, collective happiness, love, self-fulfillment, or the peace that passeth understanding) using it consistently thereafter as its principle of validation. Here it can make an infelicitous choice which will occasion conflicts in the act of validation (e.g., the selection of self-survival and love of another person are at times incompatible), just as science could sacrifice its facility for predicting if, for instance, it chose too narrow a tolerance of error. And as science is aided by mathematics, so can ethics draw upon the resources of biology, sociology, and anthropology in selecting wisely among possible principles of validation. To the details of such problems we return in a later chapter.

In his *Reason and Goodness*[12] the distinguished rationalist, my colleague Brand Blanshard, makes significant comment

[12] B. Blanshard, *Reason and Goodness,* London: G. George Allen and Unwin, Ltd., 1961.

upon the use of convention and postulate in ethics. He considers them irrational and wishes to do without them; in criticising the linguistic moralist R. M. Hare, who bases validity in ethics upon a consistent system of imperatives (much as done in this book if the principles of validation were omitted) Blanshard says:

According to this view "a moral judgment ... can be supported by arguing from its coherence with an extended system of principles, and one can say to anyone who disputes it, 'you must either accept the judgment or reject all that it implies.' This looks like an impressive advance. The trouble is that having carried us this grateful distance beyond emotivism, these moralists pull us back at the last step with an irrationalism that calls the whole advance into question. For if the imperative ... rests on a wider system, and this system in turn is a non-rational commitment, then all the commitments that depend on it are similarly in the end indefensible. ... What we have here is a free use of reason on the lower slopes with a surrender to unreasoned decision at the summit."

Unfortunately Professor Blanshard entertains an excessively exalted view of the competence of reason when he ascribes it to an ability to proceed without commitments. As we have seen there is no field of human endeavor in which the principles of reason, as Professor Blanshard conceives them, are wholly self-sufficient; even logic and mathematics are no exceptions for they involve commitments. His ideal, ethics without postulates, is wholly beyond attainment.

Perhaps this account has dwelt too long upon the arbitrary character of the decision under study. As a practical antidote to this excess let us merely note the overwhelming unanimity that exists among people in all parts of the world with respect to the validating principles of ethics. I think humanity could agree without prompting that happiness, benevolence of fellow men, and peace are among them; and with that modicum of agreement ethics, if conceived correctly as an empirical enterprise, could go a long way.

To summarize: The concerns of ethics stretch from initial commands to the area of human actions and beyond, to their validation. The commands engender specific precepts by casuistic explication; the precepts are lived by a group of men, producing determinate actions. In these actions, certain qualities, often called primary values, appear. Some of these are chosen as validating principles. If the initial specific commands produce the primary values chosen, the entire living complex is a proper, a valid, ethical system, and whatever properties the commands require human actions to possess are called ethical values. The overarching relation between imperatives and their validation creates the normative quality of values.

Thus, to argue that human life is a value in our society is to affirm the following. Our society has chosen to obey a code which includes the commandment "Thou shalt not kill." It has, furthermore, chosen as validating principles of conduct, happiness of the greatest number of people, individual freedom, and perhaps other qualities (which are difficult to specify because the choice is not consciously performed and its results are vague). To say that human life is a value (or has value), and that our ethics, which contain this imperative, are worth defending against competing systems is to affirm that our living the code has resulted in happiness, freedom, etc.

This point of view provides a method for answering all meaningful questions concerning value and virtue. Value inheres in any action required by the (validated) commands; virtue is the disposition that tends to activate such values. If the method seems cumbersome and impractical, could this not be, in part at least, because agreement with respect to the meaning of ethics has been lacking? And could not the adoption of a scientific epistemology in this subject possibly remove the uncertainties and ineptitudes that make it cumbersome and impractical?

If ethics is an empirical science, then it stands to reason that it needs a laboratory in which validation or invalidation of commands are systematically pursued. Science, in its obvious and literal implication, would therefore have us set up a group of people and require them to obey experimental commands for a period of time. After a sufficient period (maybe one or two human generations) the scientist-ethicist would then appear on the scene and determine, perhaps by psychological and sociological tests, whether certain experimental principles of validation are satisfied. If the experiment is unsuccessful he will try another combination of codes and validating principles until he has found a matching pair.

In essence this *is* indeed the typical procedure. But it is also perfectly silly. It is related to the actual pursuit of ethics as the laboratory exercise of a freshman, who uses a travelling microscope to measure the length of a footrule, is related to the practice of atomic physics. The suggestion and the proof it seeks reminds one of the attitude of people who are so pedantically "scientific" that they discount the ocean of evidence around them and demand specific proof of what is clear to all. They include the doubting Thomases who will not believe that moral indifference in the teaching of children is the cause of juvenile delinquency *unless it is proved* by psychologists in laboratory tests.

Science defies this attitude. At the present time the most highly developed natural science is astronomy. Yet the astronomer has never had a laboratory of the pedantic kind here envisioned. To be sure, the incidental and the systematic experiments of physics were of help to him in many ways, and great new insights came to him through the discoveries of physicists. But his main successes are attributable to the undeviating attention he devoted to a clearly conceived purpose, to his efforts at building a theory of celestial mechanics, to the care with which he refined his instruments. He had a laboratory, to be sure; and it was the universe.

Ethics is in many respects similar to astronomy. It can profit from controlled as well as uncontrolled experiments. It can make use of trivial examples that almost amount to controlled experimentation: Hitler's imperatives, together with all his casuistry, were lived for a while by the German people. The consequences met no conceivable standard of validation, neither happiness nor self-fulfillment nor even the most trivial standard of all, group survival. Many narrow lessons of this sort are available, and perhaps many more are needed. But of far greater significance is the ready existence of a large ethical laboratory in the form of all human history. The astronomer need not push the stars around to perfect his science; he observes their natural motions in the universe. The ethical philosopher need not push men around to perfect his "science"; he observes their natural behavior in history.

In retrospect one may wish to ask oneself whether this interpretation of ethics and of history is really so strange. I fear that sufficient reflection may even destroy the novelty of my thesis and make it appear like common sense! Certainly, insofar as ethical philosophers have attained agreement with respect to the validity of moral codes this has come about through an awareness, vague perhaps but active, that these codes have produced human satisfaction when applied to human conduct. And on the other hand, few will deny the appropriateness of the view of history which sees it as the proving grounds of moral principles. Our times need a philosophy of history; the straight-laced historian reporting what he mistakenly calls facts has overplayed his role. Injection of ethical concerns is one way of making the study of history vital, and the complementation of ethics with history, which the present view demands, clearly has advantages for both disciplines.

The approach of this book, being empirical, can provide no final solutions to the specific problems of ethics. It shows a way in which they can be solved and leaves ethical science with a faculty of improving itself as it progresses. Nor does it answer at once the question of the uniqueness of moral codes, or settle the dispute over whether one code or several lead to the satisfaction of given primary values, thereby establishing the possibility or impossibility of ethical coexistence of different codes. The immediate consequence of our considerations is to impose on us the obligation to profit from the experience under rival systems. But to these matters we shall return in later sections.

A view somewhat related to the present has been expressed by G. Lundberg,[13] who believes that the collective wisdom of mankind, accumulated through the ages, will *automatically* produce ethical behavior. If this is true, there needs to be, nevertheless, a formal study of this guidance by

[13] G. Lundberg, *Can Science Save Us?* New York: Longmans, Green and Company, 1947.

collective wisdom which is teachable and arguable, at least for the sake of speeding up the process and taking it out of the hands of chance. One may doubt, perhaps, whether the production of ethical behavior is automatic. The laws of nature, to be sure, work by themselves, the laws of ethics presuppose man's understanding and require man's active engagement in order to function.

Finally, in grateful acknowledgment, I wish to say that the writing of this chapter has profited from the work of Professor C. J. Ducasse,[14] whose fundamental conviction that ethics functions as an empirical discipline is in agreement with the content of this chapter.

[14] C. J. Ducasse, "Scientific Method in Ethics," *Philosophy and Phenomenological Research*, Vol. XIV, No. 1, September 1953, pp. 72–88

V

Northrop's Ethics

ABSTRACT

Northrop's views on ethics are similar to those presented in this book inasmuch as they also align its structure with that of science. The present chapter points out these similarities, calls attention to certain differences, and indicates where, from our standpoint, Northrop's theory invites criticism.

In several important publications F. S. C. Northrop[1] has developed an approach to ethics which is strongly oriented toward science and parallels in many respects the thesis of

---

[1] *The Logic of the Sciences and the Humanities,* New York: Macmillan, 1947; *The Nature of Concepts,* Stillwater, Oklahoma: Oklahoma A & M College, 1950; *The Complexity of Legal and Ethical Experience,* Boston: Little, Brown, 1959; Law and Morals, lecture before Council of Learned Societies symposium on law and morals (Spring, 1961), unpublished.

this book. Two significant points are there: the crux of ethics is the establishment of ("esto") norms in terms of which *de facto* sociological behavior can be judged good or bad beyond mere personal approval or disapproval, and these norms must be capable of verification, to use his preferred term. We have added to these points the demonstration that the entire formal structure of ethics is, or can be, that of an applied theoretical science, admitting however, thus perhaps deviating from Northrop's view, that the postulates of ethics are *sui generis* and have a content which is unrelated to science. Moreover, we see the starting point of ethics in deliberate commands which exact self-commitment rather than in commitment to epistemological premises.

As to the first point, the need of an *ought* which can not be derived from an *is,* we have merely echoed Northrop's forceful insistence. The problem is perhaps most clearly stated in his *Complexity of Legal and Ethical Experience,* where it arises as a question in the philosophy of law and is expressed in terms of jurisprudence. But since from our point of view man-made statutes are specific though artificial and often imperfect instances of ethical codes, the transfer from jurisprudence to ethics is a purely verbal one. With respect to the fundamental issue, he observes: "the 'ought' for judging the 'is' of a particular subject matter can not be found in the 'is' of that subject matter itself." This, he holds, is true for ethics as well as law, indeed for any discipline that transcends the factual. And he goes on to say in particular:

"Since most societies in the world today, including the United States . . . are requiring their lawyers and judges to advise and pass legal judgment on both the positive and the living law, it appears . . . that contemporary legal science must embrace more than legal positivism and sociological jurisprudence. One can no more get the criterion of the 'ought' for judging the 'is' of the living law out of the living law of sociological jurisprudence than one can get the criterion of the 'ought' for judging the 'is' of the positive law out of the 'is' of the positive law of legal positivism. In fact, there is a logical as well as a methodological block in the way." The logical difficulty lies in the fact that an actual human situation, viewed descriptively in the manner of sociology and anthropology, does not imply criteria for judging itself good or bad. The methodological block referred to is related to this: If the result of acting out a moral or legal code were to be compared, empirically, with the logical implications of the code itself, they would be found to agree "by definition," so that the act of verifying is like trivial rubber-stamping of the initial code. As the sequel will show, however, Northrop does hold that one cognitive situation (the epistemology of natural science) can provide the generic ought for the cognitive-factual situation in ethics and law.

To quote him further, and with unqualified approval: "Sociological jurisprudence falls short when lawyers and judges are forced, as is the case today, to pass judgment on the 'is' of the living law. Put positively, this means that legal science must affirm certain propositions to be true independently of and logically antecedent to both the positive and the living law.

The problem is to give methodological and objectively verifiable content to this thesis. The only factor in human experience and scientific knowledge fulfilling this condition is nature and natural man; that is, those facts in human experience, present in any society or culture, which are not the result of the beliefs of man and their deeds as directed by these beliefs."

The practical urgency of the problems under consideration appears from another passage in *The Complexity of Legal and Ethical Experience,* which I quote because of its timeliness:

"Hence, the most pressing issue confronting positive American constitutional law and liberal democratic institutions generally is at bottom a philosophical as well as a legal problem. This problem is nothing less than that of so reconstructing the theoretical and methodological assumptions of legal science that a judge of scientific integrity will be free to allow the new social legislation, without which democracy will fail, to stand, while at the same time not so tying the hands of the judge that by default of jurisdiction he will fail to protect the natural rights and civil liberties of individuals without which, also, there is neither liberty nor democracy.

The foregoing analysis suggests that a natural law jurisprudence with a content different from that of Locke and our Founding Fathers is required for such an undertaking. Even so, if this new philosophy of natural law ignores the living law of sociological jurisprudence, it will fail. Similarly, a sociological jurisprudence which does not implement itself through a reconstruction of the positive law will betray mankind also by failing to bring the available living cultural and moral resources of the world to bear upon the peaceful resolution of the disputes of men and nations. In an atomic age, such a failure is a serious thing."

Even if the ought and the is are distinct and do not entail each other on the plane of descriptive science, Northrop holds that the attainment of the ought is nonetheless a scientific matter; his last-named book ends on this note: "Evaluative as well as descriptive law and ethics are sciences." Let us now study the details of the process whereby the "ought" is generated.

The area in which one is to find them is indicated by this comment. "The ethical norms are empirically testable and therefore cognitive. The test is, however, not through their deductive or operational consequences with respect to society and culture, but through their epistemological, and other philosophical antecedents with respect to nature. It appears, therefore, that ethics and law neglect the rest of science and philosophy at their peril."

The details come in large part from a diagram in Chapter XXI of the *Logic of the Sciences and the Humanities,* which is reproduced in Chapter XXII of *The Complexity of Legal*

ETHICS AND SCIENCE

and Ethical Experience. Ethics has as its postulates or premises certain norms: our imperatives, commands or codes.

For present purposes the diagram will be simplified and a different terminology will be chosen. Let us denote the premises of a given ethical system (labelled i, i = 1, 2 . . . n) by $P^i_{ethical}$. (See right portion of Fig. 1.) From them may be deduced, by logic and casuistry, specific rules of behavior, called R^i and these, when obeyed by a group of people (a civilization, culture, nation, sect, etc.) give rise to a pattern of behavior which I shall designate as B^i. Supposedly, B^i follows by some kind of entailment from $P^i_{ethical}$ so that a comparison between them is idle and empirically insignificant.

The natural science which is contemporaneous with the ethical systems under scrutiny likewise has postulates or premises which we shall call P_{sc}. (Left part of Fig. 1.) From

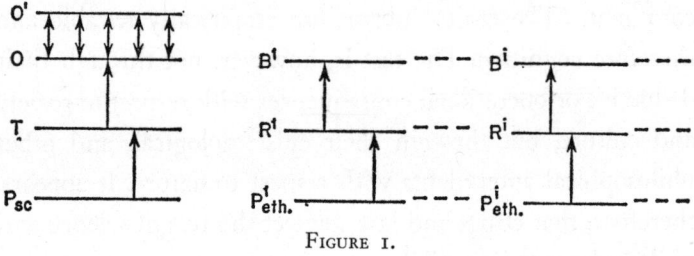

FIGURE 1.

them follow theorems of lesser logical range which will be termed T; when equipped with the rules of correspondence discussed in our Chapter I they lead to a predicted set of observable phenomena O. This state of affairs is graphically symbolized in the figure below. Note the complete analogy between $P_{ethical}$ and P_{sc}, R and T, B and O. However, there exists above O a level of *actual*, not merely predicted observations, which carries the symbol O'; the prime is meant to accent the *de facto* character of the observations. Hence a comparison between O and O' is possible and this comparison is not tautological or rubber-stamping. Similar non-implied *de facto* levels do not occur above B, and this means that verification is not possible in any ethical system. Realizing this, Northrop proposes that comparisons be made between the various $P^i_{ethical}$ and P_{sc} at the very basic level of postulates and that correctness be attributed to the ethical system whose $P^i_{ethical}$ agrees with P_{sc}, evidently supposing that a judgment as to closest agreement is feasible.

So much for the architecture of his arguments. A closer study is best conducted in connection with examples which he himself proposes. Let ethical system number 1 (Northrop's synonym is "normative social theory") be what he calls the patriarchal theory of heredity, meaning a social situation in which family ties are strong, actions are based on feelings of filial affection and other unconstructed immediacies that are close to the P-plane of Chapter I. Its premises, $P^1_{ethical}$, are said to be the epistemological contentions of naive realism, and one is left to suppose that R^1 is something like the body of precepts found in Taoism and in patriarchal societies, while B^1 is the behavior displayed by their members.

Ethical system number 2 is described in all its particulars. It is a combination of the ethics of Confucius, Buddhism, and Hinduism which, according to Northrop, dwell primarily upon the P-plane and are compatible with the moral philosophies of the Stoa and of Locke and Jefferson; the latter, operating largely in the C-field of Chapter I, are held responsible to a considerable extent for the American constitution. The union of these types of ethics into a proper marriage is unobjectionable, because as we have seen, concepts given in intuition, i.e., P-facts, are connected with concepts by postulation or constructs by "epistemic correlations," which is Northrop's name for our rules of correspondence. I think that this synthesis of the systems of East and West[2] is admirably ingenious and suggestive.

The premises of system 2, $P^2_{ethical}$, are decomposed into three separate propositions to which I shall here refer as p_1, p_2 and p_3; that is to say $P^2_{ethical} = (p_1, p_2$ and $p_3)$, which is the conjunction of p_1, p_2 and p_3. To wit:

- p_1 affirms that a person in his primary (P-plane) awareness is not only his differentiated self but also his indeterminate field consciousness of himself.
- p_2 affirms that a person, like every object, is also a construct.
- p_3 affirms that p_1 and p_2 are compatible, that because of this compatibility man can merge or fuse his sensory and his rational being into one complete personality.

[2] See Northrop, *The Meeting of East and West*, New York: Macmillan, 1947.

NORTHROP'S ETHICS

There is doubtless a trace of panpsychism, perhaps of the Oriental variety, contained in p_1. As to the other two partial premises, p_2 is in agreement with Chapter I, while p_3 again invokes the blessing upon a marriage made possible by the rules of correspondence.

The crucial demonstration, leading to the conclusion that ethical system 2 is preferable to system 1, that 2 is indeed the correct system, involves two parts. One is to show that (p_1, p_2, p_3), the conjunction of these component premises, does in fact generate ethical precepts with *esto* or ought character (by being first recognized as cognitively true and then soliciting existential commitment). The other is to confirm the agreement between (p_1, p_2, p_3) and the premises of the methodology of modern natural science. I am persuaded that Northrop has succeeded in this latter task but believe that the former is impossible. To give substance to this belief, I shall go into particulars.

With reference to Fig. (1), the points just made may be restated as follows. Agreement between $P^{(2)}_{ethical}$ i.e., (p_1, p_2, p_3) and P_{sc} exists. But to rise from $P^{(2)}_{ethical}$ as defined above to any sort of $R^{(2)}$ presents peculiar difficulties.

We first focus attention on p_1. This proposition, we are told, "gives the thesis that all human beings and all other natural objects in their intrinsic nature (as distinct from their accidental sensuous *esse est percipi* effects upon the observer) are not merely equal, but also identical." One might be inclined to ask here how this conclusion follows strictly from man's field consciousness. For suppose we admit the existence of an indiscriminate field consciousness, devoid of all distinctions between you and me. Then at this stage no *logical* conclusion can be drawn at all, in particular a statement alleging the identity of you and me cannot be made because there is no you and me. If the logic of inference is to be employed, one must pass from p_1 to p_2, thereby establishing the constructs you and me. But then the identity is destroyed, a contrast has been opened in the C-field, and the ethical connection between persons must be sought in other ways, independently of arguments based upon the lack of clear distinctions between you and me in a primitive stage of awareness. Here, in fact, all the traditional problems of ethics reappear.

On the other hand, Northrop's phrasing of p_1 also permits an interpretation which places the identity between you and me in the domain of the "undifferentiated continuum" which should perhaps be called the "barely" or vaguely differentiated continuum. If it is adopted the conclusion is even more difficult to secure because, surely, the effacing of important conceptual differences in aesthetic awareness, the blending of objects into a continuum which occurs in a meditative mood cannot be sufficient reasons for claiming them to be identical. Nor does it seem plausible to assume that P-experience completely wipes out the I-thou distinction.

But after all the author does not say that the conclusion is entailed by p_1. He carefully uses the words "gives the thesis" and does not appeal to logic. It may be, therefore, that the preceding criticism misses his point, that the thesis follows from p_1 in a more pragmatic or mystic sense, perhaps in the way Francis of Assisi was induced to speak of "my sister the cow" and couple this relation with ethical content. In another place, Northrop suggests that a Japanese painter who wants to paint bamboo must identify himself with a bamboo stalk before he is able to paint it. Such suggestions make a great deal of sense to me, for I have a special fondness for Japanese painting, but I would hesitate to establish an ethical relation between the painter and the bamboo on these grounds. At any rate, let us observe that, if the conclusion: all human beings are identical, is accepted—whether it be derived from p_1 or not—there is as yet no *ought* prescribing human behavior.

To establish this, Northrop falls back upon p_2 which, he says, "gives the following criterion of the ethically good and legally just." Here he introduces a rather complicated set of definitions and stipulations, bearing upon four further propositions (labelled [i] to [iv]) which I shall state as clearly as I am able to do. The matter is important because here is where the ought rabbit comes out of the scientific hat.

Let p be a person, x a normative predicate (good, bad, virtuous, etc.),[3] s a factual situation in conduct or action (like

[3] Northrop actually calls x a normative judgment. If he means by this a full proposition containing both subject and predicate, like "stealing is bad," and by s a specific action like "stealing a watermelon," the critique given here

stealing). Then the statement, s is x, (e.g. stealing is bad) is morally good, i.e., carries an obligation, if the four conditions listed below are satisfied. To avoid repetition, I shall comment on each of them individually after it is stated.

(i) x is *in accord with* a universal law involving constructs and holding for all p.

Now it is certainly not a matter of simple logic to show how a normative predicate, like "is good," can be in accord with a law. That is not even possible for an ordinary predicate, e.g., the property of being blue, for a law bears upon sentences, not predicates. Yet if x is expanded into a full sentence or judgment, the situation is not much better so far as normative sentences are concerned. Then, of course, there is no difficulty with descriptive or factual sentences—"certain compounds of cobalt are blue" is, in a very primitive sense, in accord with scientific law. But a normative sentence, "this action is good, or ought to be performed," can never be an instance of any law as the word is used in science. What it means for a normative statement to be in accord with a law is therefore problematic. Were we to accept Northrop's reasoning at this point on other grounds, for instance by saying that the normative "good" must be interpreted as "good for

---

would be slightly different but would yield the same effective conclusion. In that case, however, his phrase, "to say s of x is morally good," is difficult to fill with meaning because s is then contained in x, and the statement is a tautology which can not be morally good or bad. For this reason I suppose here that x is a predicate.

all men," and the universal law as "all men want to live"[4] —we should be guilty of smuggling in the ought surreptitiously, and indeed without logical warrant; there is no reason why the ethical person may not insist that there are conditions under which men *ought not* to live. To repeat: the ought can not be caught in the web of science, nor can it be pulled out of it. Our own procedure, which builds it into the ethical premises at the beginning by making them commands, is not embarrassed by this circumstance. However, it, too, remains incomplete and useless unless it is supplemented by validations, as we have seen.

The second condition reads:

(ii) If the action s confers specific rights, privileges and duties upon one p, these must be accorded to all p.

In terms of an example, if s means stealing and stealing is compatible with a universal law (which, if our criticism of (i) is accepted, it can not be) then everybody is permitted to steal. This is clearly an ethical postulate, synthetic and rising far above any implications or quasi-implications of condition (i). As an axiomatic adjunction of this sort it is quite meaningful, and it resembles Kant's categorical imperative, whose nature and shortcomings are discussed elsewhere in this book (see p. 208 et seq.). Notice, however, that this interpretation of condition (ii), proper though it may be in a different

---

[4] This interpretation, which I believe to be the one Northrop intended, would place his theory very close to Kant's. For a review and criticism of Kant's categorical imperative see Chapter VI.

scheme of ethics, encumbers the neatness of Fig. 1, which is a condensation of Northrop's diagram. The passage from level P_{ethical} to R is no longer simple and analytic; it involves the injection of extraneous, indeed normative, assumptions from the outside, thereby further reducing the symmetry between the ethical and the scientific state of affairs.

The third condition demands that

(iii) The law in question, which according to (ii) is valid for all p, as well as the proposition "for all p, s,"[5] must be freely accepted by implicit or explicit consent of the parties concerned.

This pronouncement again is, and is meant to be, postulational, for we are told that the law (which is true for all p) "being postulationally and contractually constructed and hence merely hypothetical," requires condition (iii) as a supplement. When thus understood, the present item becomes a logically unconnected but wholly acceptable declaration against authoritarian ethics. Since it is postulational, it is subject to the criticism under (ii). The problem here under discussion is again considered in Law and Morals. There, as elsewhere, the ladder leading from the cognitively true to the normatively obliging has one additional rung. We are asked to ascend from first-order facts to the recognition of the truth of "second-order facts" and from there by existential self-commitment to the ought. The interpolation of second-

---

[5] Literally, if s is a specific action, that is hard to understand. I take it to mean: (p) x ⊃ s. Example: for all people, stealing is bad.

order facts, if indeed they are facts of the cognitively true or false variety, does not relieve the need for normative postulation.

Condition (iv) reads (verbatim):

(iv) "There are the concepts by formal imageless postulation of a logical (as distinct from a naive) realistic epistemology." This, as I see it, represents a reiteration of an important point already contained in premise p_2 itself.

We have commented at some length on the way in which p_2 "gives," to use Northrop's term, a criterion for the ethically good. The considerations above show that the simple word "gives" is not to be construed as "implies," since some of the conditions p_2 (i) = p_2 (iv) are postulational and do not "explain" the esto character of ethical laws. Undeniable, however, is the fact that the commitments suggested by the separate parts of Northrop's creed are found to be harmonious, and are found together in the living habits of real Western people. This symbiosis, according to the thesis of this book, has its explanation in the circumstance that the common parts of the codes of Buddhism, Hinduism, Confucianism and Stoicism (and perhaps Locke's philosophy whose importance in this connection is not quite clear to me) as well as the less imageful, constructive ethical systems of Paul, Augustine, Aquinas, Spinoza, and Kant which are based on a conscious recognition of the role of rational constructs, have in fact been sufficiently validated in terms of various accepted principles of validation—such as survival, happiness, human perfection, etc.—to be regarded as viable and as binding in historical perspective. Literally, they are a *tradition* because of their successful appeal to living men or, if you are a pragmatist, their utility.

Let us now return to Northrop's systematic theory, waiving the preceding strictures, and observe it in action. P_{sc} in Fig. 1 represents the epistemology of current exact science, especially physics which, through the work of Einstein, Heisenberg, Schrödinger and Dirac, has taken on an increasingly abstract and constructively rational complexion, departing more widely than ever from naive realism. But $P^1_{ethical}$ represents naive realism, whereas $P^2_{ethical}$ is identical with P_{sc}. Therefore, ethical system 2 is to be accepted, while 1 must be rejected.

From my point of view, two blemishes are present in this theory of ethics. First is a certain lack of symmetry between the scientific and the ethical parts of Fig. 1. The scientific structure has an additional level O′ at the top, a level which makes empirical confirmation possible. In consequence, scientific verification must occur at the *top,* ethical validation at the *bottom* of the methodological structure. Here is an unfortunate imbalance which destroys all facilities for treating ethics in the manner of a science. Second, the ethical scheme, for any given system, is not self-contained, is not complete; for the passage from $P_{ethical}$ to R, which generates the ought, necessarily relies on extraneous postulates not included in $P_{ethical}$.

NORTHROP'S ETHICS

Symmetry can be restored, and the scheme can be made complete, by the following changes, which form the gist of Chapters I–IV. Install the nucleus of an "ought" in $P_{ethical}$ by frankly making $p_1, p_2 \ldots$ *commands* and not relating their content to the epistemology of science. Comparison at the bottom is then precluded, but the rise to rules or precepts having suasive power, R in the diagram, is now logically possible. Finally, to complete the symmetry, add another level B' above B, an ideal level defined by our principle (or principles) of validation. This is done in Fig. 2.

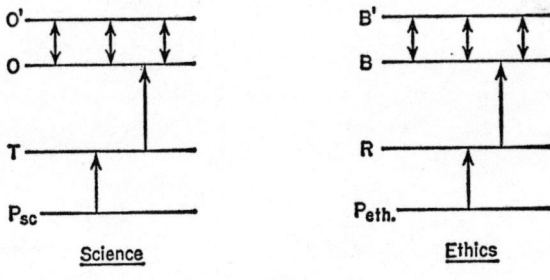

FIGURE 2.

The latter do not appear in Northrop's scheme, where they are not needed because the emergence of the normative claim of propositions is kept obscure. Their inclusion confers an attractive parallelism upon the formal methodologies of science and ethics: $P_{ethical}$ is the analog of P_{sc}, R of T, B of O, and B' of O'. The adjective formal, however, should be stressed. In content, the partners of these pairs differ radically.

$P_{ethical}$ is composed of imperatives, P_{sc} of declarative sentences about wholly different matters; R is prescriptive, T again descriptive; B and O are both matters for observational report and therefore more nearly alike than the other levels; finally B' and O', though both serve as standards for comparison (for B and O), exhibit a curious reversal of roles: O' is *actual*, observed behavior in the scientific world whereas B' is an *ideal* state defined by principles of validation. I do not think that a closer, material correspondence between ethics and science has been attained or is worth striving for, since it clearly does not exist even between the various natural sciences themselves.

VI

The Historical Confusion between Validating Principles and Ethical Imperatives

ABSTRACT

Primary values, while vaguely connected with "human nature," cannot be derived from any kind of knowledge. Theories of moral emotions are unable to account for them. Nor are ethical imperatives derivable from other principles, not even from the primary values. The relation between imperatives and primary values is empirical; only the symbiosis of men in history, inspired by the imperatives, can enact the primary values. It can also fail to enact them.

Moral genius sometimes establishes an intuitive connection between primary values and imperatives in a manner akin to the inductive leap of science. The confusion between the two, coupled with a belief that one necessarily entails the other, is responsible for the common view that ethical codes can not be validated.

THE POSTULATIONAL CHARACTER OF PRIMARY VALUES

Principles of validation or, succinctly, primary values were likened in Chapter IV to the conventions which enable science to verify its theories. The latter rests upon overwhelmingly secure pragmatic foundations but cannot be derived deductively from any independent propositions whatsoever. We wish to show that the same is true for all ethical principles of validation.

Two of them will concern us above others, partly because they present the features of interest here in concentrated essence and partly because they have played dominant roles in ethical thinking. I am referring to hedonism (or, in its wider meaning, to eudemonism and utilitarianism) and to Kant's categorical imperative. Introducing the latter in this context seems at first erroneous, for its very name indicates that it is a command and not a principle of validation, nor is there any question about its author's meaning and intention. Kant took it as a command and placed it at the beginning, not at the end of the ethical enterprise. We show, however, that its name is mistaken, that it cannot function as an imperative but only as a primary value in the sense explained, although even as such it requires augmentation.

Hedonism is a systematization of human aims in terms of happiness, either individual or collective. To be sure, it often adds to its substance the trappings of commands, admonishing us to act in ways that promote happiness and pleasure, or that minimize unhappiness and pain. More elliptically, the doctrine speaks of the rightness of an action and fixes it in proportion to the pleasure it engenders or the pain it avoids. At this point I prefer to strip hedonism of its pretense to admonish and take it as nothing more than a statement of human aims, i.e., as a primary value. In the next section the problem is to be raised concerning whether, and by what principles, one can properly choose the actions that lead to happiness, and we shall conclude that the problem is either trivial or requires an empirical solution, a solution which does not flow merely from the statement that happiness is a human aim.

Here we ask: Why the aim? There are those who regard hedonism as *self-evident,* as conveying in its very essence the intuitive assurance that it must be right. It is difficult to argue with them, for they are unimpressed by the universal catastrophe which has swept away all self-evident synthetic truths from logic, mathematics, and natural science. And I am unwilling to grant it an asylum in ethics. If the demand for consistency and uniformity throughout all intellectual areas has no appeal to the defender of hedonism, then one must point to instances which *contradict* happiness and pleasure as human aims, with the reminder that a truth, if self-evident, cannot be violated. But there have been men who preferred to *suffer* for a noble purpose. Throughout the ages one encounters groups which practiced self-denial, self-mortification as a method to attain ethical perfection. In contrast to hedonism, asceticism is the note of every great religious reformer, it is heard in the appeals of Moses, Gautama, Paul, Manu, Mohammed, St. Francis and Luther. These are men whose ethical testimony cannot be brushed aside as irrelevant or mistaken; the argument that they were deluded or ignorant, as were the people who denied certain other supposedly self-evident propositions in science (such as the axioms of Euclid) because they failed to understand them, can hardly be applied here. On the contrary, asceticism is a *rival* to hedonism; the contribution of individuals from Moses to Luther is to be seen in the same light as the discovery of non-Euclidean geometries, which broke the spell of the doctrine of self-evidence once and for all.

We have made two points concerning the self-evidence of hedonism: one, that modern developments in large areas of thinking suggest a fallacy in every claim to self-evidence, and two, that there are in fact ethical philosophies contradictory to and competing with hedonism which were and are held by persons of supreme ethical wisdom. For both these reasons we abandon the view that hedonism escapes being a postulate because it is self-evident.

If it is not self-evident, nor a postulate, then it must follow logically from some anterior, more basic proposition. Could it be derived, perhaps, from the universal nature of Man? If to be a man necessitates the quest for happiness, then hedonism is proved. But there are unhappy men; indeed, to use a phrase which is not wholly clear in its meaning, there is perhaps more unhappiness than happiness in the world, and it is probably easier to show that unhappiness is the human lot than its reverse. This, however, is somewhat beside the point, for it is not relevant to our argument that man *be* happy; it is essential to show that he craves happiness. To show that this is often false let me refer back to the proponents of asceticism already named, to the cynics who deliberately renounced the quest for pleasure, riches, honors and all the agreeable things of life, to the Christian hermits who shunned pleasure as an impediment to the life of grace. I see no way in which the thesis can be made plausible that man's necessary and essential state is one of seeking happiness.

Psychologists have sometimes said that the motive of every action is to bring about a condition more strongly accented by pleasure than the present. This statement makes the claim of hedonism true by definition. Unfortunately, however, it is of no help in this connection, for aside from entailing hedonism, it also defines pleasure, and it does this in a most peculiar way. What ordinarily counts for pain is turned by definition into pleasure, for instance by a monk who, in the act of flagellation, maintains in this act that satisfaction of the soul outweighs the body's distress. Or else it is alleged that momentary pain which results from a present action is to be subordinated to ultimate pleasure in the future. These maneuverings are redundant and ineffectual; they fall short of establishing the positive proposition that search for happiness is inherent in the nature of man.

Jeremiah Bentley and John Stuart Mill, the chief advocates of utilitarianism in modern times, did not believe its aim to be a simple naturalistic fact. The first of them takes it very nearly as a postulate. After he raises the question as to a direct proof of the principle of utility[1] he tends to answer it in the negative, saying:

"That which is used to prove everything else, cannot itself be proved: a chain of proofs must have their commencement somewhere."

Mill,[2] on the other hand, while admitting that the principle is not susceptible to direct proof "in the ordinary and popular meaning of the term," presents considerations which he regards as equivalent to proof. "The utilitarian doctrine," he says, "is that happiness is desirable, and the only thing desirable, as an end; all other things being only desirable as means to that end. What ought to be required of this doctrine —what conditions is it requisite that this doctrine should fulfill—to make good its claim to be believed? The only proof capable of being given that an object is visible, is that people actually see it. The only proof that a sound is audible, is that people hear it; and so of the other sources of our ex-

[1] J. Bentham, *An Introduction to the Principles of Morals and Legislation*, Oxford Press, 1879.
[2] J. S. Mill, *Utilitarianism*, London: Longmans Green & Co., 1888.

perience. In like manner, I apprehend, the sole evidence it is possible to produce that anything is desirable, is that people do actually desire it."

Here is one of the classical instances of fallacious reasoning, often exposed by Mill's opponents[3] in the past. Despite their similar endings, the words visible, audible and desirable refer to different classes of experience. Visible and audible denote what *can* be seen and heard, but desirable does not mean what *can* be desired; it means what *should* be desired. Therefore a factual demonstration can establish what is visible and audible, but not what is desirable. The latter requires precisely the kind of entailment by some more fundamental and more embracive insight we have been seeking.

Thus far our chief concern has been with a form of hedonism, individual or personal hedonism, which most students regard as an insufficient basis for ethics. For reasonable use that form must be expanded; instead of striving for one's own greatest happiness, one must seek maximum happiness among all people. Let us not worry at this point about the awesome difficulty connected with the operational meaning of a phrase like maximum happiness among all people, but accept it for the moment as well defined.[4] Even then the expansion poses problems that are insoluble. Mill admits that

---

[3] See for instance, E. Westermarck, *Ethical Relativity*, New York: Harcourt Brace, 1932.

[4] A promising attempt at a quantitative definition of "satisfaction" has been made by N. Rashevski in his *Mathematical Biology of Social Behavior*, Chicago: University of Chicago Press, 1957.

no reason can be given why the general happiness (not only of all mankind but) of the "whole sentient creation" is desirable, "except that each person, so far as he believes it to be attainable, desires his own happiness. This, however, being a fact, we have not only all the proof which the case admits of, but all which it is possible to require, that happiness is good: that each person's happiness is a good to that person, and the general happiness, therefore, a good to the aggregate of all persons."

The transition from the aim of individual to that of collective or universal happiness, allegedly demonstrated as cogent by this argument, involves a nonsequitur, which is rendered more difficult to detect because of the diversionary injection of the undefined and irrelevant idea of "good." Westermarck (*loc. cit.*) uncovers it, saying: "If a person desires his own happiness, and if what he desires is desirable, in the sense that he *ought* to desire it, the standard of general happiness can only mean that each person ought to desire his own happiness. In other words, the premises in Mill's argument would lead to egoistic hedonism, not to utilitarianism or universalistic hedonism."

The hiatus between egoistic and collective hedonism is also clearly seen by Sidgwick,[5] who tries to bridge the gap by introducing an axiom or principle of rational benevolence, i.e., the proposition "that each one is morally bound to regard the good of any other individual as much as his own."

[5] H. Sidgwick, *The Methods of Ethics,* London: Macmillan, 1874.

He calls this, rightly, an axiom but bases it on intuition, holding it to be no less evident than mathematical axioms. Here we see a correct appraisal of the logical status of the utilitarian aim, even though in this avowal utilitarianism, trying to establish itself, undergoes a metamorphosis into the gentler ethical doctrine of benevolence.

There are other forms of hedonism, among them a most persuasive one advanced by Wilmon H. Sheldon.[6] As is well known, this sensitive and circumspect philosopher bases his arguments on deep religious convictions and makes no claims with respect to any distinguished logical status of the doctrine. Hence his view cannot be contradicted. He is certainly right in saying: "The supposed refutations of hedonism are the most superficial arguments in all the history of ethics. They are like an attempt to disprove the law of gravitation because many bodies are prevented from falling when we lift them up." But he would likewise admit, I think, that hedonism cannot be *proved* in the manner of Bentham, Mill and others who have attempted to do so. To him it seems to be a pragmatic principle which engages the best resources of human reason, feeling and aspiration. Moreover, in his magnificent defense of hedonism Sheldon goes on to show how crude desire is altruistically ennobled and how happiness, thus purified, can serve as a primary value in terms of which our actions can be judged.

Enough has been said to indicate that the aim of hedo-

[6] W. H. Sheldon, *Rational Religion*, New York: Philosophical Library, 1962.

nism cannot be established as a natural fact, nor is it something that can be shown to be necessary or inevitable in the face of other cogent facts or considerations: it is a postulate, an aim men may decide to set themselves, a star by which they may wish to guide their lives, a criterion they choose to invoke when judging the successes or failures of their actions, individual or collective. It is a principle designed to validate the imperatives, which are the forces generating actions.

We now turn to Kant's ethical philosophy, assessing it, too, from the viewpoint of epistemology: what, we ask, is the ground for the truth claim in his universal moral law? That "law," we recall, commits us to acting "always according to maxims which thou canst wish to be accepted as universal laws." A maxim, as distinct from a moral law, is understood by Kant as a mere rule of conduct based upon experience and, therefore, subject to opinion, variable and not binding upon all. What it means to act in accordance with a universal law, in what manner the existence or indeed the knowledge of a universal law inspires or defines specific individual actions, these are questions reserved for detailed discussion in the following section. Here our interest is in the goal to which our actions are said to be dedicated: the acknowledgment and the behavioral support of a universal law for all mankind. This is still rather vague: to delineate Kant's conjecture clearly I find no alternative to supposing that he desired as the ethical goal a situation in which all men can live together; and if I am pressed for final particulars, I would have to add: live happily, contentedly, live in fulfillment of their human potentialities, live in freedom, or, at the very least, *survive*. That is to say, I am really confessing that my reading of Kant has not convinced me of his success in providing a well defined, clearly stated goal when he formulated his universal moral law. And when his aim is thus supplemented it differs little from those of utilitarianism of the various kinds. This verdict does not damage Kant's system beyond repair, for the strength of his position rests, not in the statement of his goal, but in the rationale which surrounds it; not in his acquiescent acceptance of eudemonism but in his analysis of its relation to the *good will* and to *duty*. These most important points of Kant's ethics do not concern us in the present context.

To round out the picture, other forms of the basic law which its author thought to be equivalent will also be cited. At another place in the "Metaphysics of Morals" one reads: "Act so as to use mankind, both in thine own person and the persons of others, ever as an end, never merely as a means," and again: "Act according to the ideal will of all rational beings, as the source of a universal legislation." Let us note here first that these statements, though couched in the form of imperatives, are no true imperatives at all. They point to a state of affairs whose prevailing justifies human actions but never specifies these actions. Hence they are, or claim to be, principles of validation rather than imperatives.

Each of them, however, requires again the sort of augmentation in terms of *qualities* that characterize the state of affairs which universal legislation is intended to achieve, precisely as was set forth in connection with the first version of the categorical imperative in the foregoing paragraph.

Kant saw very clearly that his "imperative" could not be derived from experience. Reason, supreme and independent, "the author of her own principles," is the source of it. Here, in the ethical domain, Kant's system attains that completeness, that bilateral symmetry which make it so impressive. Pure reason, active in the domain of thinking (in science), engenders the categories which outline, guide and fashion theoretical truth, and it does so without the aid of datal experience. Practical reason, active in the domain of willing (in ethics), gives rise to the ideal law here under consideration, and does so without reliance on moral experience. Indeed Kant holds that the *good will,* good only because of its orientation to the moral law, is defiled by considerations of happiness, prudence or any purposive foresight whatsoever. Ethics differs from science insofar as practical reason differs from pure reason, but each creates the logical foundations upon which its elaborate edifice, ethics for one and science for the other, can be erected.

Kant did not regard his moral law as a postulate, and therefore his account of its origin differs from that here maintained. The difference is related to the history of the idea of absolute truth which was recorded in Chapter I, wherein Kant occupies a curious but interesting place. In the beginning of Western philosophy, absolute truth has its locus primarily in the axioms (originally *aitemata*) of logic, science and religion, whose credibility was vouchsafed by divine inspiration. Divine inspiration was replaced, as time went on, by some sort of innate evidence present within the axioms themselves, and theories arose which purported to explain this evidence in terms of man's *lumen naturale* or by a process of reliable intuition. During these early phases absolute truth retained its character of certainty, immutability and adequacy to all experienced fact, but its genesis remained a mystery. Then Kant appeared on the philosophic scene. Displeased at the mysterious origin of absolute truth but still convinced of its reliability and its changelessness, he constructed a dual mechanism for producing it and called this mechanism reason, pure and practical, with gears called pure forms of intuition and categories, duty and moral imperatives. This cumbersome apparatus, he felt, supplied a proof for the validity of axioms—indeed for the whole wide class of propositions known as synthetic judgments *a priori* —so that his major problem was solved: Kant felt he understood both the eternal validity and the origin of axioms.

But he could not foresee the end of the story about absolute truth, or, in case we are not yet at the end, the developments that took place since his time. All major branches of inquiry have come to the conclusion that, somehow, if the eternal validity of axioms is maintained, one cannot understand their origin, and conversely, if one tries to understand their origin, one cannot maintain their eternal validity. This complementarity has been resolved by rejecting eternal validity, that is, the very concept of absolute truth, and by simultaneously refining our understanding of the genesis of axioms. Nor is this an arbitrary resolution, for there are numerous historical instances in which reputedly "eternal truths" have revealed their falsity. Hence the former axioms have become postulates, posits, useful premises for deduction which can change if and when experience requires changes. Kant stands in the middle of this historic process. He is separated from Aristotle and St. Thomas by about the same distance as we from him. But he was sensitive to the demands of the science of his day, and one is safe in assuming, I believe, that he would regard his moral law as a postulate in the modern sense were he alive today.

In setting forth the postulational character of primary values, of our principles for validating ethical norms, we have dwelt upon hedonism, utilitarianism and upon Kant's moral law (or categorical imperative). It is unnecessary, I think, to go through all the other ethical doctrines. Take the aim of self-fulfillment. Assuming that it can be defined, that I fully understand my natural endowments and capabilities, I can still ask why I should fulfill them. Logically, it is as plausible that I should curb my natural propensities, as it is to indulge or cultivate them; hence the postulational nature of the maxim of self-fulfillment is apparent. The same is

HISTORICAL CONFUSION

true for Sorokin's mandate of altruistic love, and for every kind of evolutionary ethics, especially for the latter. For if I am to behave in accordance with the evolutionary destiny of the human race—provided even that I know that destiny—and wish to render account of this obligation beyond its acceptance as a posit, I must at least show the falsity of those doctrines which take nature to be evil, take the evolutionary process to be ethically crude and in need of correction. I do not see how this can possibly be done.

Special mention must be made of one primary value, namely, survival, either of the ethical person or the ethical group. While rudimentary, it comes closer to being derivable from valid anterior considerations than the others and carries a powerful measure of inherent evidence. For the principle of survival, if violated, defeats the ethical process itself and thus makes ethics meaningless; it has, therefore, been argued that survival is a necessary or an obvious primary value. But is it? There is nothing *logically* wrong with the statement that mankind is basically evil and ought to obliterate itself, nor is this impossible, as it should be if the principle were not a postulate.

There is an interesting interpretation of the religious doctrine of the "fall to sin," which takes the dilemma into which man has fallen to be this. Man is unable to see from his human situation what his goal should be. This interpretation reflects in a way the point we have here tried to make.

Theories of Moral Emotions

Those who do recognize that primary values are neither derivative nor self-evident facts, but who fail to see the need and the usefulness of postulates in every human context, either regard primary values as unique and intractible by ordinary methods (G. E. Moore, R. Hartman) or as expressions of subjective emotions (Adam Smith, Brentano, Westermarck, Ayer, Stevens). The latter interpretation requires comment because it is in one sense very close to ours and still it seems to arrive at vastly different conclusions. If our *reasoning* about the world involved only facts interspersed with self-evident truths—and this was the view widely held until the beginning of the twentieth century—then values clearly do not fall into the domain of reasoning. Now the other categories of mental activity admitted by classical psychology were perceiving, willing and feeling. The first of these has obviously nothing to do with values; the second has been called upon to supply the basis of ethics (Fichte, Schelling, Schopenhauer) but is no longer very widely accepted; the third has served and is still serving as a favorite philosophy under the name of "emotive theory of values."

One form of that theory (Westermarck's) maintains that there are two specific moral emotions, complex in nature but nevertheless distinguishable from others, called *moral approval* and *moral disapproval* or indignation, and these two alone have led to the formation of concepts like good and

HISTORICAL CONFUSION

bad, right and wrong. This latter distinction, whose detailed analysis in terms of the emotive theory we here forego, reveals the character of these two sentiments as moral emotions. Nevertheless, both of them belong to a wider class of feelings which Westermarck calls retributive emotions. These, by their nature, relate the person having the emotion to other persons who are the causes of the moral approval or disapproval. Thus, the moral emotion of approval is closely allied to, and generates, certain non-moral retributive emotions ranging from kindly feelings to gratitude, while the moral emotion of disapproval is similarly associated with another class of non-moral retributive emotions like anger, hatred and revenge.

The technical difficulties which this theory encounters are numerous; they occur especially when an attempt is made to separate the specifically moral emotions from others. The fact that actions often, if not always, spring from feelings cannot be doubted; but whether these feelings are the primary causes of action or are induced by subtler ideas, whether in case they are primary, only certain emotions (belonging to a narrow class) lead to moral actions—these are questions to which the answers are equivocal. McDougall,[7] for example, denies both primacy and uniqueness of moral emotions, saying that "judgment of approval may be prompted by admiration, gratitude, positive self-feeling,

[7] W. McDougall, *An Introduction to Social Psychology,* London: Methuen & Son, 1908.

or by any one of the emotions when induced by way of the primitive sympathetic reaction; judgment of disapproval springs most frequently from anger, either in its primary uncomplicated form, or as an element in one of its secondary combinations, such as shame, reproach, scorn, but also from fear and disgust." Yet if all this is granted, if indeed there are no *specific* moral emotions and one must, therefore, assume that all emotions have moral potency, the theory may still survive these strictures: Ethical values then become vague expressions of rather general multiform feelings.

Further evidence accrues to it from an observation which its proponents are always eager to cite: The relativity of ethical judgments. Westermarck's writings[8] emphasize the variability of human behavior in different civilizations; he surveys in interesting fashion the killing of parents worn out with age or disease, the exposure of children, attitudes with respect to suicide, homosexuality, and normal sexual practices. If ethics were based on objective and verifiable judgments, he holds, such variations would have to be written off as errors of judgment, but the errors would be astoundingly universal. The theory of moral emotions, on the other hand, accounts for differences in moral valuation in terms of the different situations and external conditions of life which influence emotions.

From our point of view, such references to *de facto* be-

[8] Cf. especially E. Westermarck, *The Origin and Development of Moral Ideas*, London: Macmillan, 1906.

havior are irrelevant. We have already made a clear distinction between *est* and *esto* norms; sociological behavior and customs concern the former and give, when taken by themselves, no significant clues to *esto* norms.

Aside from this there are more important reasons for rejecting theories based on moral emotions. One of them emerges when an account is demanded of how emotions relate themselves to *judgments,* and there *are* moral judgments even if they are protested as meaningless by the emotive school. Practically, they are the most important judgments of our time. A person or a nation can have two kinds of enemies, those who hold emotional dislikes and those who disapprove on grounds of moral judgments. Now it seems clear that the former can be swayed by arguments and acts designed to alter feelings, but I doubt very much whether the latter enemies can be talked out of their bellicose attitude by an insistence that their disapproval is meaningless. The *need* for a positive, rational, indeed scientific criterion is paramount. These comments do not directly support the thesis that such a criterion *exists;* they are meant to accentuate its importance if it does exist. The *existence* of moral judgments is clear and present, and is not denied by the emotive school.

One may look upon feelings as interposed between judgment and action. Every fully premeditated act has as its antecedent a rational judgment and this usually releases feelings, prior to the act, which make the latter seem pleasant or distasteful (but not right or wrong), make us do it from desire or from duty. There can be feelings without rational antecedents, feelings without issue in action, and feelings lacking both. None of these, I think, gives rise to *responsible* ethical phenomena; in case they do lead to action, that action is impulsive (and therefore non-ethical), otherwise the situation is merely emotive. To say that we control our feelings, that we repress them or give them free reign, is to affirm their subjection to rational principles. If this interpretation is correct, it implies that feelings cannot serve as starting points for an ethical theory.

This state of affairs is not improved by the attempt to change the sequence: judgment-feeling-action, and to allege that feelings *create* judgments which result in action. To lend credence to this point, the Germans have invented an appealing word, *Gefühlsevidenz,* which smoothly suggests that emotions bear rational evidence. In this I can see nothing but a trick, an amusing play on words.

The foregoing remarks are based on acceptance of a distinction between reasoning and feeling and argue in favor of selecting the former as the source of primary values. Can this distinction be maintained? The answer to this question deals, I think, the most serious blow to the doctrine of moral emotions, although it also dissolves the previous objections by making them unnecessary: Modern psychology finds it difficult to maintain the old categorical distinction between concepts and emotions. Every proposition, to use Wundt's phrase, is "gefühlsbetont," and practically every emotion is associated with conceptual knowledge: reason and feeling are inseparable as psychological activities. Hence a theory which invokes a clean division between these faculties is unconvincing.

Another way of saying this is: Ethics should not be given a psychologistic start. It is better to proceed in terms of such logical concepts as postulation, implication, deduction, confirmation and so forth. This is my own belief; yet I hasten to add that it is not quite as strong as that of the proponents of logical universalism who believe in logic as the cure of all our ills. Aside from the results it promises, my preference for the logical approach comes from the realization that the logical terms, which are also not susceptible to absolutely precise definition, are nevertheless clearer than the psychological ones. But it is not denied that there is also a psychology of ethics, important in its own right, though not likely to add cogency to the acceptance of primary values.

How, then, were we launched upon this excursion into psychology? Let us recall. Primary values, our principles of validation for ethical imperatives, were found beyond rational proof. They entered the scene as postulates. This, to some, is a shocking development, enough to send them scurrying off into the countryside for evidence from feelings, ready to abandon what is sometimes conceived to be the scientific approach.

Listen again to Westermarck.[9] After expressing frustration at the impossibility of establishing primary values by rational means, he says: "If there are no moral truths it cannot be the object of a science of ethics to lay down rules for human conduct, since the aim of all science is the discovery of some truth." It is as simple as that! Compare this now with the more elaborate and (apologies for the boastful claim!) more appropriate version of the scientific method given in Chapter I, and the need for re-evaluation becomes obvious. Science is not the discovery of "truths"; it is an accommodation of factual experience by models or constructs, and by its very nature has need for unprovable postulates. Some of these, as we have seen, are required at the beginning, others, serving the purpose of verification, come at the end. Commitments to primary values are like the latter. They are *not* ruled out by science; and their very presence in ethical theory creates strong presumptive evidence suggesting that the structure of ethics is very much like the structure of science.

The proponents of moral sentiment theories argue that there is no objectivity in ethics. Such would be the case if objectivity were to be found only in factual certainty or *a priori* knowledge. In science, however, objectivity lies in a certain conformance between theoretical models and protocol facts, and if this view be carried into ethics, the problem of objectivity is open, not to be settled in so simple a way. But I

[9] E. Westermarck, *Ethical Relativity*, New York: Harcourt Brace, 1932.

HISTORICAL CONFUSION

doubt if the problem of objectivity, in the scientific sense of *objective reference* to some protocol experience, is important in ethics. What matters crucially is the possibility of obtaining *universal agreement* to an ethical system and this is a different problem altogether. At this point the parallel with science may not be drawn too closely.

THE CONNECTION BETWEEN PRIMARY VALUES AND COMMANDS

One of the greatest fallacies perpetuated by Western moral philosophers is the thesis that *knowledge* of the good, and by this is meant an approval of certain primary values, implies imperatives for moral action. It began with Socrates' extreme belief that wisdom is goodness, and that he who knows the truth will do the good. If truth means correct and adequate knowledge of the kind conveyed by science, then as we have shown, there is no road from truth to ethical behavior; what *is* cannot prescribe what *ought* to be. Even if truth or wisdom includes an avowal of primary values, as Socrates supposed, there is still no certain way of inferring how the wise man must act in concrete circumstances; the primary values do not define an ethical code. Before marshalling the evidence for this observation, let us note that the Socratean position is foreign to Oriental thought, especially to Buddhism, which emphatically declares its ethical norms to be the quintessence of past experience, the wisdom of the ages accumulated by saints who lived these norms, and not by contemplation of what is true, or what it means to be good. And the Tao, with all its seemingly mystic implications, was "the ancient way."

The point I wish to make in this section is so simple as to be almost obvious: To know whether you attain the ethical goal to which you are committed you must live by certain precepts and see whether these precepts lead you to the goal. These precepts may be vaguely suggested by the goal but are never implied or fixed by it, for it is humanly impossible to know in non-empirical fashion what course of action leads to happiness, survival, self-fulfillment, achievement of evolutionary goals or the peace that passeth all understanding. This is evident to anyone who has pondered over the relation of entailment between precept and promise of Christ's "Blessed are the meek for they shall inherit the earth."

Let us see in detail what sort of directives can be spun out of the doctrine of hedonism. Personal hedonism, the striving for one's own happiness, is not very interesting here because there are few people who believe that it forms a satisfactory basis for ethics. Those who do must provide supplementary restraints upon egotistic behavior, and this sometimes involves an appeal to love as a universal, natural humanitarian attitude, as with Wilmon H. Sheldon. If it is to stand by itself, hedonism is satisfactory only in its collective form, which aspires to universal happiness. How does one go from there to specific rules of action? Is stealing, for instance, prohibited by the goal of collective happiness?

To covet my neighbor's bank account is a natural desire

for me if I am penniless. To steal some of it would cause me happiness at the expense of his displeasure. Hence, if the doctrine in question is taken literally, the act of stealing is permitted if the amount of pleasure I derive from the act outweighs the amount of his displeasure. At present, it is true, the bookkeeping of happiness presents serious problems, yet it is conceivable, though doubtful, that psychology will some day establish objective measures of pleasure and pain which will permit the accounting here envisaged. For the sake of argument, assume this to be the case. We can then establish an injunction against certain kinds of overt stealing that lead to detection and result in unhappiness of the victim greater than the thief's pleasure. Even if I, myself, were not found out, the loss experienced by the other person might overbalance my anonymous enrichment. And if this situation is arithmetized, a million such acts will cause more pain than pleasure, which rules them out.

The commandment, "Thou shalt not steal," however, is not understood in this limited sense. It also prohibits the embezzlement of public funds whose disappearance goes undetected or, if detected, causes only a minor ripple of unhappiness among the administrators of the funds. It outlaws stealing from a rich man who would hardly feel the loss. One might say, of course, that if every poor person stole from the same rich man, the latter would experience serious displeasure; to prevent this, the hedonistic principle must require that the rich man's negative pleasure is greater than the sum total of the pleasures of all the thieves. Even so, it could not rule out the morality of procedures in which each poor man steals from a different rich man. In other words, hedonism contradicts the usual interpretation of the seventh commandment by sanctioning acts which that interpretation prohibits.

Such arguments are well known, and are frequently used to indict hedonism. But this is fair only if hedonism claims the competence to regulate detailed behavior. There are factors in the interaction among people, in their disposition to mercy as well as to legalistic retribution, in their fallibility and their ignorance, which these arguments entirely ignore, and which could not be included in the most elaborate accounting of happiness, even if psychological measures for that elusive quality were available. The passage from the goal to maxims of behavior is not a strict logical transition; it cannot be performed with sufficient compulsion in that direction at all. There is a possible passage which will be discussed in the following section under the title of the "Inductive Leap"; it is more like a flight than a logical movement. The reverse transition, from maxims or commands to goals, is not strict and logical either, but it can be performed and is being performed continually in the transactions of men, in the process of history. It is an empirical movement from imperatives to values, and one whose outcome cannot be foreseen theoretically. To take it unanalyzed, to rely on it implicitly, is to accept "the ancient way."

HISTORICAL CONFUSION

Our detailed illustration dealt with the prohibition of stealing, and it tried, but failed, to deduce it from the hedonistic postulate. The troubles we encountered arise whenever that postulate is squeezed to yield imperatives. Thus it seems to make lying an ethical act if it allows me to escape punishment for a wicked deed, for it is hardly true that my confession of the truth will make a policeman happy. Honesty in general is difficult to defend. And if any doubt is left in the reader's mind about the inconclusiveness of efforts to deduce concrete rules of action from hedonistic principles, he might try to justify legal punishment, especially capital punishment in their name. Volumes have been written on this theme, and we shall not elaborate it here.

The Golden Rule is an imperative which seems to spring directly from the principle of universal hedonism. Simply put it says: Treat your neighbor as you wish him to treat you. This sounds plausible enough and appears to contradict our claim that primary values do not entail imperatives. A moment's thought will show, however, that the Golden Rule, in spite of the good intentions of its advocates, is not a valid ethical imperative at all.

Consider healthy competition for worthwhile ends, which most people will regard as ethically desirable. The Golden Rule does not permit it, for if I want to get ahead of my competitor, I must let him get ahead of me. If I like to smoke, must I then induce my neighbor to smoke? Zeno, Seneca, Cato desired death as their suicides showed; were they therefore entitled to kill their friends? Here, as in many similar situations, the Golden Rule is indiscriminate with respect to special situations. Nor does it say anything about actions without direct effects on others (be neat, perfect yourself, seek an education!). Last, but not least serious, is the fact that it prohibits punishment for immoral acts.

Kant's categorical imperative hardly makes the task of getting rules for action any easier. The philosopher himself is not very helpful here, for his main concern throughout his writings (*Grundlegung zur Metaphysik der Sitten, Kritik der Praktischen Vernunft, Einführung zur Tugendlehre*) is to assure and convince his readers of the majesty of his moral law, of its utter aloofness from the claims of hedonism, utilitarianism, and eudemonism. Yet its content remains nebulous, a phantom in the sky spelling DUTY, but lacking indications as to duty of what.

Kant's feeble attempts to show that honesty is enjoined by his categorical imperative have never been convincing. I may find it convenient and proper to lie or steal under the unique circumstances of the moment, and it may be perfectly rational for me to will the maxim of my dishonest act into a universal law which should regulate the will of all, *under these unusual circumstances*. Indeed I can always regard my special circumstances so unique that, in the parlance of modern mathematics, the instances of their recurrence form a set of measure zero, so that the elevation of my dishonest will to the level of a universal law will be innocuous, permitting me to lie while requiring everybody else to speak the truth. Since every set of circumstances is unique and unusual, this premise has no effect upon the conclusion, which is that dishonesty is not implied by the categorical imperative. But the same result follows even without the premise. There is no logical reason why deliberate lying cannot be willed into a universal law: The group doing so will probably be unhappy, neurotic or die out, all of which is all right so far as the categorical imperative is concerned; for that law has deliberately been so emasculated as to bear no reference to matters that are emotive and utilitarian.

We need not belabor these difficulties, for they are too well known. They were, in fact, the substance of the first and most incisive criticisms levelled against Kant's moral arguments. Hegel[10] says: "A principle that is suitable for universal legislation already presupposes content. . . . The criterion that there should be no contradiction produces nothing." Westermarck[11] summarizes the thoughts of many critics when he reasons: "Even in the cases which Kant has carefully selected to demonstrate his principle, he cannot help considering human nature, social conditions, and the consequences of acts, however much he deprecates a morality that depends on experience. Lotze and others have remarked that without a consideration of consequences almost any maxim might be suitable to be presented as a universal rule;

[10] E. G. Hegel, *The Philosophy of Right*, London: G. Bell & Sons, 1896.
[11] E. Westermarck, *Ethical Relativity*, New York: Harcourt Brace, 1932.

it is its consequences that decide whether it is suitable or not. . . . It has been said that the weakness of his imperative of duty is that it 'lacks all organic filling,' that it is an empty form without contents, an unconditional command which commands nothing."

As we have seen, the categorical imperative is not an imperative at all; nor is it strictly speaking among the primary values, the principles of validation. However it does take the form of one of these if it is filled with content, e.g., by specifying that the intended universal law is one which allows humanity to survive, to progress in certain ways, to be free, to be happy and so on. But even then, as we have seen, it generates no duties precise enough to form the basis for concrete actions.

Here, then, lies the great historical confusion with regard to the methodology of ethics. Too frequently has it been assumed that the commands are implied by the values. In fact, however, imperatives and primary values are independent variables. Both must be posited, and the process of living must validate the postulation and establish *empirical* compatibility between the two, must show that the values validate the norms (imperatives), and that the norms are adequate to the values.

The parallelism between this conclusion and the method of science, as outlined in Chapter I, is now apparent. In science, instead of wanting to survive or be happy or be free, we wish to predict observable phenomena within a certain

HISTORICAL CONFUSION

range of error. This is an end freely chosen—perhaps ultimately for ethical reasons. No one can say, without appeal to experiment and observation, whether a given theory, i.e., a set of constructs derived from the postulates made at the beginning of its logical range, is compatible with the purpose it is designed to attain: prediction of protocol facts. Figures 1, a and b illustrate the parallelism between science and ethics and suggest the manner in which ethics, like science, can be regarded as an empirical discipline. They are simplified and annotated versions of Fig. 2 in Chapter V.

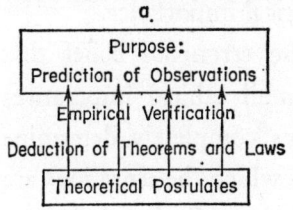

FIGURE 1.

Science	Ethics
begins with theoretical postulates, then moves by deduction of specific laws and theorems toward its final, postulated purpose: Prediction of Observations.	begins with ethical commands, then moves by explication (casuistry) and actual living in accordance with the commands toward its final, postulated purpose, attainment of primary values.

229

The Intuitive Connection between Commands and Primary Values

In the effort of showing that primary values do not strictly imply commands, I have deliberately remained short of saying that there is no connection at all. The empirical development which goes, or fails to go, from commands to goals through the process of living the commands, has already been recognized. There is a second connection, based on intuition enlightened by experience, which is akin to the inductive leap in science (cf. Chapter I, page 42) and demands attention because of its historical importance.

The scientific counterpart of the erroneous belief that principles of validation strictly entail ethical imperatives is the Lockean view that sensations completely determine the constructs or models in terms of which the sensations are to be explained, determine them in the radical and elementary sense in which constructs (Locke's complex ideas) are mere combinations of P-experiences (simple ideas). However plausible in Locke's day, this extreme empiricist philosophy can no longer be maintained. While it is true that Galileo's theory: all bodies fall with constant acceleration, is very nearly suggested by observations, the Einstein metric, from which the observations follow by a long series of mathematical steps, certainly is not. Nor do the ideas of the carbon cycle "follow" from the data of astronomy, nor the Mendelian characters from experiments on peas, the isotopic spin from the observed regularities of the periodic table, the quantum theory of resonant bonds from observations on chemical valence, nor the Freudian *id* from the overt behavior of men. The systematic connection between facts and constructs is not reciprocal; as the developments in Chapter I have indicated, there is a unique passage from the latter to the former: Einstein's metric leads uniquely via the process of deduction to theorems, thence to specific scientific propositions and finally, by way of rules of correspondence, to observable facts. Yet the observed facts may be compatible with other initial postulates which are perhaps not known. It is this latter circumstance which allows for the changes in fundamental conceptions on which the progress of science vitally depends.

Nevertheless we know that fundamental concepts spring up vis-à-vis the facts, are "suggested" by them, as the saying goes. Their emergence, in a sense, sets aside the rules of deductive inference; it involves the inductive leap which amounts to an ingenious guess followed by systematic exploitation of its consequences. This is most clearly seen in the *abstract* sciences, where the facts offer a minimum of suggestive guidance; in the *descriptive* sciences, where explanations are simple and fairly obvious, the point here made is often obscure or trivial.

If ethics were a simple science like Galileo's theory of falling bodies, the foregoing remarks would be uninteresting. The step from observation to the law of constant acceleration is very short and looks like pure induction. Similarly, in simple forms of ethics like individualistic hedonism, the actions that make a person happy are practically dictated by the goal—not entirely, perhaps, because an untried course of action may leave me in doubt as to its revenue in pleasure —but by and large I can foresee the happiness which will result. Modern ethics, however, is a much more complex structure and its correct analog is an abstract theory of science like relativity, where the facts fail to suggest the concepts. Here it is the inductive leap which takes the ethically creative person from the primary values to the commands, and the living of men in history under the commands fills the gap by proceeding in the opposite direction.

Moral genius establishes the commands by performing the leap, usually without specifying the primary values which inspired these commands, often even confusing the primary values with the commands. Mankind then lives these imperatives, and when the style of their lives displays the primary values, the imperatives are validated, the commands then acquire normative qualities, and the good becomes defined in greater detail than was the general goal set by the primary values alone.

VII

Origins of Imperatives and Primary Values

ABSTRACT

Most ethical systems contain admixtures of non-ethical considerations. This fact is natural, for the methodology of ethics is "open-ended," so that metaphysics and religion can render important aid in supplying material for choice as imperatives and principles of validation. Brief comments from this point of view are made on: The Egyptian Cult of Osiris, Mazdaism, Judaism, the ethics of Lao Tse, Confucius and Gautama. Christian ethics and the anomalies that attach to Paul's doctrine of love are singled out for more extended discussion.

The Induction of Ethical Attitudes by Non-Ethical Doctrines

Most ethical systems with historical efficacy, especially in the West, contain notable admixtures of non-ethical considerations. This is natural and inevitable, for the pure methodology of ethics as outlined in this book is somewhat anemic and, when standing alone, is forced to rest its case upon a degree of philosophic maturity in its practitioners which cannot be counted upon among the people whose commitment to imperatives is necessary to make the system work. Nor is the system opposed to or incompatible with certain metaphysical and religious convictions. While in principle it does not need them, it may nevertheless draw support from extra-ethical beliefs.

The fact is that ethical attitudes are often *induced* by religious and metaphysical considerations. Our theory does not imply which of these, the ethical or the non-ethical component, is more important, nor does it establish an irreversible causal relation between them. Moral philosophy traditionally indulges in an oversimplification: having seized upon ethics as the central issue under study it sometimes tends to regard all others as secondary, as merely contributing to the main concern and as significant only in that role; or, in the other extreme, having seen a frequent historical connection between religion and ethics, moral philosophy sees them as related through necessity. In fact both components are important in human experience, not by virtue of an invariable causal relation—which does not exist—but because living and thinking are complexes of many-termed relations, and the terms are accented differently by different persons and in different cultures.

To support these claims, let me recall once more the occurrence, especially in the Orient, of ethical systems which do not derive sustenance from religious or metaphysical beliefs in our sense. Hence ethics do not necessarily generate religious attitudes. And to discredit the theory that religious beliefs, when present, invariably induce ethical norms by way of irreversible logical entailment, let us note in passing that a well known philosophic doctrine reverses this very claim and attempts to infer the existence of God from the assumed ethical goodness of men. More to the point, perhaps, is the fact that early religion frequently did not have any true moral bearing. Breastead,[1] speaking of the Babylonians, points out that "offences were merely ceremonial transgressions against the god. . . . The penitential psalms clearly indicate that the evil consequences, which the offender so passionately pleads he may escape, are due not to disapproval of evil conduct by the god, but, as Westermarck had noted, to the curses of the injured party." And Westermarck[2] observes that "in none of the penitential psalms

---

[1] J. H. Breastead, *The Dawn of Conscience*, New York: Scribners' Sons, 1933.
[2] E. Westermarck, *Origin and Development of Moral Ideals*, vol. II, London: Macmillan, 1906.

known to us is there any indication that the notion of sin comprised offenses against fellow men."

Induction is always a reciprocal process, as even its simplest example in physics indicates. Since the term is drawn from physics, the reader will permit me a parenthetical illustration from that science. In the process called electrostatic induction the separation of charges on one body induces a potential on another; but a potential on one body may also induce a separation of charges on another. In an analogous way religion can induce ethics, and ethics religion. An outstanding example of the latter process will be given later in this chapter (page 250 et seq.).

How, then, does the induction of ethics by religion (or metaphysics) fit into our formal scheme? We have seen that this scheme is doubly open-ended, that it requires a choice of imperatives at its start and of principles of validation at its terminus. These choices can be made very effectively on the basis of religion and metaphysics. The latter supply sometimes the imperatives, sometimes the validating principles, and usually a mixture of both.

The Egyptian Cult of Osiris. To illustrate, let us consider the polytheism of ancient Egypt, which (possibly because our knowledge of Egyptian culture is somewhat scant) shows forth the connection in its simplest form. There is the legend of Osiris who, having been assassinated by Seth, now rules the empire of the dead. Pharao is his messenger on earth and is charged with dispensing the divine justice of Osiris.

At death, man's soul appears before Osiris and is required to justify its conduct during its life on earth in order to be granted immortality. And here the "Book of the Dead" spells out the *imperatives* which must be observed if immortality is to be gained: Cheat neither men nor gods; do not lie; do not suppress the workers, or harm the poor, the weak, the slaves or the children; help those who are in need; obey the laws of the king.

This rationale, which couples morality with trascendental rewards, is typical of many Western religions and appears with variations and refinements in Judaism, Christianity and Islam. It yields directly the imperatives. Whether it specifies primary values is not quite so clear; if it is thought to do so in promising immortality, then it turns out that this primary value cannot function as a principle for *empirical* validation, because immortality cannot be attained in this life. Further scrutiny, however, reveals a different state of affairs which is compatible with the empirical aspects of our methodology.

Clearly, the attainment of immortality in an after-life cannot serve as a principle of validation; the *expectation* or *assurance* of immortality during this life, however, can! Such blissful assurance serves as a simple and rather primitive criterion, and probably one to be subsumed under the larger classification of happiness, security or mental tranquility, to which it contributes.

If it is accepted, religion can be seen to furnish in this instance both the imperatives and the primary values required by the ethical system; the submersion of ethics in the matrix of religion closes both ends of our outline and accomplishes what would otherwise be left to human choice. He who rejects the expectation of immortality as a criterion of validation can only grant that the religion of Egypt supplies the imperatives but not the primary values, and he must look elsewhere for a principle of validation.

This example shows what advantages accrue from a coupling of ethics and religion. The disadvantages, on the other hand, are equally clear. The *un*believer, having been taught that the coupling is obvious or necessary, will upon rejecting this belief surrender his ethics along with his religion and be cast adrift without norms. This outcome is too disastrous and too general in history to be dismissed casually. It accentuates the need, in principle, of an ethical theory like the present, in which the postulational open ends of ethics *can* be closed in *non*-religious ways, just as they were closed in science. The Egyptians, incidentally, seem to have lacked this insight; we are told in some heretic documents that, because nobody ever returns from the empire of the dead, compliance with the code in chapter 125 of the Book of the Dead is nonsense, and one might as well seek happiness by indulgence in the present life.

ORIGINS OF IMPERATIVES AND PRIMARY VALUES

Mazdaism. Mazdaism in ancient Persia provides a more dynamic interplay between religious doctrine and ethical precepts inasmuch as it integrates man into the historical process and holds him responsible for the fate of the entire cosmos. It engages man in a fight under the leadership of the good god Mazda against the corruption of the world wrought by Ahriman, who represents the principle of evil. While according to the Zend Avesta, the sacred book of Iran, the final outcome of this struggle is predestined to be favorable to Mazda, man's participation in it is nonetheless essential; each of his acts can influence the fight. There evolves in this context a code of good behavior, strangely similar to all other Western codes, which prescribes honesty and justice and establishes the obligation to work, in addition to various duties peculiar to the members of an agricultural society, duties which today we would regard as social rather than ethical. There is also the promise of life after death in the heaven of Mazda for those who have battled to advance his cause; and the unique contribution of Mazdaism to ethics seems to be its emphasis upon human responsibility for the universal course of events. In a sense, this emphasis anticipates Aristotle's principle of active self-fulfillment and foreshadows the ethical evolutionisms of our day.

Evidently the metaphysical background of this doctrine engenders both, imperatives and principle of validation, although the latter is again none too clear. One might equate it to the development of man's capacity for action conducive to Mazda's victory, but there is also evidence that Persians strove for happiness, either in this life or after death.

Judaism. In early Judaism[3] we encounter a religious doctrine which produced clearly formulated and impressive norms, the Mosaic code, but a rather crude principle of validation. *Jahwe* is a righteous, indeed a vengeful god, who will not tolerate other gods before him, will reward those who obey his commands in this life, but visit punishment upon those who break them and upon their children to the fourth generation. Jewish monotheism must be acknowledged as an important and probably a new phase of religious thinking; still one is left to wonder whether the conception of a single god was not entailed by the Hebrew instinct for authoritarian ethics. If God's jealousy, his vengefulness are to guarantee man's good behavior, then God cannot tolerate competition in his own sphere. From that point of view the postulate of a single god is a concession to ethical needs rather than a lofty metaphysical conception.[4]

Notice that the reward for good behavior comes in this life, either in the present or in future generations. The prin-

---

[3] Judaism, like Christianity and most other religions, later evolved into a bewildering complex of philosophic and ethical views which render its fundamental theses increasingly amorphous and obscure.

[4] Indeed the literal history of the Jehovah concept bears this out. The spirit with whom Jacob wrestled at the brook Jabbok was doubtless a local god, an "el," as evidenced by the fact that he gave Jacob his name, Isra-el. Moses replaced the "el" by "Yahve," the name of a local god of a desert tribe. Growing reverence later caused the praying Hebrews to pronounce the consonants of Yahve with the vowels of the word for "Lord," and thus resulted the name Jehovah.

ciple of validation is therefore an empirical one, directly applicable by living men. While the Old Testament contains references to life after death, to Heaven and Hell and to angels, it is almost certain that these ideas are later importations into Judaism from other cultures.

Thus we observe here, in better perspective than almost anywhere else, the quasi-historical genesis of ethical imperatives in the story of Mt. Sinai, and the announcement of the validating principle in the form of God's favor toward men.

The Anomaly of Christian Ethics: Paul

The religious content of Christianity, though subject to conflicting interpretations when viewed in sectarian detail, is fairly homogeneous and unique. It is primarily a doctrine of human redemption through divine grace and as such answers a universal craving of all sensitive men for relief from the oppressive state of impotence before evil, before fate and before sin. In this important respect it is an extension and a refinement of the messiah idea which developed during the later stages of Judaism. Added to it are the beliefs in a life after death; in a benevolent and merciful[5] rather than a wrathful God, often the belief in eternal punishment or reward and, to lend supreme authority to the doctrine, in some form of the divinity of Christ.

[5] The Old Testament, too, occasionally emphasizes benevolence; see, for example, Isaiah.

But the ethical core of Christianity is less solid and less homogeneous. There is a substantial kernel of Judaism in the concept of duty toward God the father; the legend of Osiris survives in the prospect of heaven, parts of Mazdaism in the struggle between God and the Devil in which man's good deeds are enlisted. The most distinctive contribution, however, and the one extolled as the noblest expression of the human heart, is St. Paul's doctrine of love and charity. "If I spoke the language of men and of angels,—if I had the gift of prophecy, if I knew all the mysteries . . . if I surrendered all my goods to the poor . . . and had no love, it would avail me nothing." Or again, (1 Cor., 12) "And now abideth faith, hope and love, these three; but the greatest of these is love."

In this context of Paulinism, even Christ's sacrificial death loses some of its redemptive quality and becomes an exemplar of altruistic love, an ethical symbol for man to espouse. Man's actions, according to this precept, must be bathed in indiscriminate love toward all fellow men, and it is expected that this love will generate the psychological climate in which actions automatically become ethical. The disposition of the mind, not the nature or the purpose of an act, bestows moral rightness upon behavior; it is the spirit, not the deed that counts.

Our civilization has bowed to the delicate and sentimentally engaging qualities of this ethical doctrine of love in spite of the fact that it stands in considerable contrast to oriental and even some earlier Western doctrines, like the systematic indifference of Stoicism, which are espoused by a greater portion of mankind than is Paulinism.

Two things can at once be said about the law of love, love in its proper sense, quite apart from its human appeal which, though quite real, is in my view not primarily ethical. First, it alone can never be successful. To love everybody anonymously first and then act is simply an impossible demand. It is natural for a man to love his children, his wife, his relatives and friends, and it is equally natural to act benevolently toward them. But love is something which can hardly be artificially created, created without specific cause; this fact is partly acknowledged and partly confounded in the voluminous writings about the difference between *agape* and *eros*.[6] The failure of the thesis that love must precede every ethical act—and I think it has failed in human history—is therefore understandable.

Paul's ethics defies the formalism this book tries to establish. If it is satisfactory as a model, our contention is wrong.

---

[6] The original Greek words for love were *philia* and *eros*. Neither satisfied the sense of Paulinist ethics. Hence the early Christian writers chose the word *agape* (used previously by the philosopher Philodemus to denote a meal eaten by a group in fraternal spirit) which has been translated as *caritas, charity, compassion, brotherly love, Nächstenliebe*. Some of these portray psychological hybrids, artificial emotions incapable of effective existence. Brotherly love cannot be anonymous. Charity, on the other hand, is not at all what Paul needs as a prerequisite for his ethics, because it is induced by seeing others suffer—the German equivalent is *Mitleid*—while ethical actions should be independent of the state of suffering of those whom the action affects. For an extended discussion of this and related matters see A. Nyrgen, *Eros and Agape*.

Now I question the model as *ethics*—not necessarily in its character as a desirable state for people of certain temperaments and endowments—because it is a hopeless one. Even if it were true that people imbued with indiscriminate love act ethically, and if furthermore universal love were a humanly possible quality, its attainment as a pre-condition for ethical behavior might be so difficult as to destroy the workability of the ethical enterprise from the very beginning. The history of Paulinism seems to bear this out.

In addition, the model has the drawback of discouraging positive but painful moral acts which spring from certain conceptions of duty or even from prudence and caution. And finally, is there not a sense in which it offends the dignity of man by alleging that he will act properly only when his mind is harmoniously composed to make him *want* to do so, but never from a feeling of *obligation?* Does the doctrine not use charity as a means to achieve ethical ends as one uses a drug to evoke certain actions? Does it not impair man's responsibility? Perhaps Nietzsche's violent reaction to what he called the *Sklavenmoral* was not entirely unwarranted.

I have said that this aspect of Christian ethics violates the general scheme here proposed. This is, of course, because it sets no imperatives—except in the vaguest forms such as "love thy neighbor"—and it espouses no principle of validation, no primary value useful as a test. This is recognized and well phrased by Blanshard (*loc. cit.*) who says: "Love contains in itself no principle of selection. Unless guided by

ORIGINS OF IMPERATIVES AND PRIMARY VALUES

an insight other than its own, it showers itself alike on the just and on the unjust, and is capable of passionate devotion to personal and communal ideals that reflection can only condemn."

There is a more favorable way of looking at the problem. One may regard love or charity not as a direct command, not even as a generator of imperatives, but as the principle of validation itself, as the goal to be reached. Understood in this way the system would demand: Act in a manner that will cause all men to love you, and in response you will then love all men. This to me seems meaningful, but it reverses cause and effect in Paulinism. And even then, as we have seen in Chapter VI, the doctrine lacks the imperatives which it needs to start. For what it means, in terms of specific modes of actions, to induce love in others can only be settled by the empirical process of living.[7]

The *practical* difference between the appeal to love as a

[7] Our account here is quite incomplete. It should describe the philosophy of Max Scheeler who speaks of the "primacy of love" and uses this concept to establish a hierarchy of values. It should include other doctrines of total surrender—not love—which labor under the same difficulties as the insistence on love. One of these, which uses sentimental detachment to transform man's being, is found in the teachings of some stoics. Among them is the view that all virtues are one; man cannot be partially virtuous. This attitude has also sometimes (Marcus Aurelius) led to the denial of importance to any vigorous action, to the conclusion that it matters little what you do to your fellow man so long as it does not disturb your equanimity, or as in Seneca, to an advocacy of suicide: *Quid erubescitis? Patet exitus!* Further echoes are heard in Pascal and in Schopenhauer, who wishes to base ethics on pity.

The attempt to base ethics on love is not entirely confined to the Occident. It is present in the Eastern appeal to *mahā* and *karunā* and speaks eloquently in our day through the writings of Zen Buddhists like Suzuki. (See *New Knowledge of Human Values,* Ed. Maslow, New York: Harper's, 1958).

generator of ethical imperatives and its use as a principle of validation is doubtless not as great as our *theoretical* analysis makes it appear. For even if love comes at the end of the validating chain, one need not wait for its establishment. Ethical conduct can be induced by attention to every part of the total ethical transaction, and if love, which is normally the result of moral actions, can be taught or engendered by other means and then results in moral action, such teaching promotes ethics. A similar point can, of course, be made about all other validating principles. Happiness, for example, is widely regarded as one of these. Strictly, this means that we act morally *in order* to attain happiness. Yet if it is true that happy people are more likely to act morally than unhappy ones, then any means which makes men happy serves ethical ends.

Our whole analysis here deals of course with the *ethical* aspect of *agape*. The purely *religious* doctrine, the contention that man's love for his fellows results as a response to God's unbounded selfless love for all men quite apart from all moral considerations, is left deliberately aside.

Many readers may feel that the foregoing appraisal of Christian ethics was unsympathetic and indeed mistaken. I plead guilty, in a degree, to this charge, admitting now that the account was written to make a point. Love in its true and primary sense can not function as a precondition for ethical behavior, and the ineffectual sweetness characterizing a good deal of Christian moralizing is occasioned by a fail-

ORIGINS OF IMPERATIVES AND PRIMARY VALUES

ure to see this point. Hence the preceding. Thoughtful Christians, whose views will now be sketched, take a different stand.

They loosen the meaning of the word love and define as "agape" an attitude quite different from ordinary love, so different that the word love becomes misleading. Love, in this latter version, becomes impersonal and ceases largely to be an emotion. I have not seen this extension of meaning drawn more clearly and defended more fervently than in the recent work by my revered senior colleague, Wilmon Sheldon. In his *Rational Religion*[8] he says "Intellect itself would not function if it didn't love knowledge." He establishes the difference between liking and loving in a way that makes love ethically potent almost by fiat. I quote an important passage:

> So far then we have the maxim: put yourself in his place, her place, its place. And while that is not so hard to do with those we like, it is very hard to do with those whom we dislike, abhor, abominate. Here swims into view a most important distinction: the distinction between liking and loving. In liking persons or animals we would seek their welfare because we want them to be with us, to contribute to our pleasure by their agreeable qualities, things we enjoy witnessing. . . . Notice too that we like inanimate things: food, fresh air, clothes, furniture. Often we say we love them. . . . So

[8] Wilmon H. Sheldon, *Rational Religion*, New York: Philosophical Library, 1962.

do we confuse liking and loving, as if love were only a more intense liking. . . . On the other hand, true love is of the person or animal just for *his own* good, no thought at the moment of the pleasure that may give to ourselves. So we may love those whom we dislike, even detest. To love them is to seek their true welfare as just theirs. Not, of course, that this always means to seek their *immediate* happiness. To love your enemies needn't mean to invite them to dinner; it may mean to see them punished for their cruelties, yet as a means to their ultimate reform and moral progress. *And it is reason, and reason alone, which can justify love in this its true sense.*

There can certainly be no doubt about the ethical competence of love if that concept is so greatly expanded. And when thus accepted it invites as well as survives the methodological analysis of the preceding chapters: a loving person becomes an ethical person, almost in our sense.

Another approach to this enlarged ethics of love is afforded by Sheldon's theory of compassion and repentance, which agrees with that of most other Christian writers. Repentance, he says, is an instance of love. Here, then are the details of the workings of love. A misdeed causes suffering on my neighbor's part and thus compassion, *Mitleid,* in myself. This is repentance. To avoid this Mitleid ("suffering with") I must not injure. Kindly deeds, on the other hand, cause gratitude and benevolence on my neighbor's part, and this induces pleasure in myself. The positive disposition which prevents a man from committing acts that would evoke in him compassion, while urging him to kindle by his acts in all others happiness and gratitude, which reflect pleasure into him, is thus called Christian love.

One of the most magnificent testimonials to this concept is the ending of Goethe's Faust.

We said that this is ethics "almost" in the sense here advocated. The reservation springs from an insufficiency still manifest in this version of love. For it makes an affirmation rather similar to the thesis of emotive theories of ethics and is therefore slightly vulnerable to the arguments put forth against them. Remorse and repentance cannot be wholly explained as feelings of sympathy for others, for they accompany acts which are known to be sinful only by the individual who committed them. The *pater peccavi* confession is an ethically genuine and normal reaction to transgression even if it actually gives pain to the unsuspecting father who is asked to forgive. Love of the father, coupled with the desire to spare him sorrow, is not an excuse for witholding the confession. Here, as in other instances, ethical rigor transcends the demands of love and clings to duty, the duty imposed by valid imperatives.

ETHICS AND DIVINE MERCY: LUTHER

In a certain sense Paul's doctrine of love effects a separation between ethics and the divine-reward types of religion, for it affirms that moral conduct constitutes no claim upon

God. This view was developed into a rather extreme position by Luther who went at times so far as to say that good works without faith (rather than Paul's love) are "idle, damnable sins." And he admonishes his friend Melanchthon, who is morosely exercised with scruples of conscience over peccadillos, to sin lustily ("pecca fortiter"). This aspect of Lutheranism, exaggerated to be sure in these selected fragments from Luther's none-too-systematic writings, represents the closest approach in Western culture toward an indigenous religion that is uncoupled from ethics.

Because of this disengagement one, and I believe the primary, voice of religion speaks here in its clearest tone. Man *knows* his ethical imperatives and, in a unique experience which is utterly foreign to classical philosophy with its conviction that knowledge is virtue, finds it beyond his power to fulfill these commands. Despairing of his goodness, frustrated before his inability to satisfy what he knows to be his duties on other grounds, he needs, seeks, and experiences the existence of an infinite reservoir of mercy and calls it a God of grace. Unable to find the strength for moral rectitude he craves forgiveness, which no ethical consideration can bestow.

Roland Bainton[9] relates this insight, which animates Luther's teachings, to an early discovery by the reformer concerning the meaning of "God's Justice." In his words,

---

[9] Roland H. Bainton, *Here I Stand*, New York: Abingdon-Cokesbury Press, 1950.

"the study of the apostle Paul proved of inestimable value to Luther and at the same time confronted him with the final stumbling block because Paul unequivocally speaks of the justice of God. At the very expression Luther trembled. Yet he persisted in grappling with Paul, who plainly had agonized over precisely his problem and had found a solution. Light broke at last through the examination of exact shades of meaning in the Greek language. One understands why Luther could never join those who discarded the humanist tools of scholarship. In the Greek of the Pauline epistles the word "justice" has a double sense, rendered in English by "justice" and "justification." The former is a strict enforcement of the law, as when a judge pronounces the appropriate sentence. Justification is a process of the sort which sometimes takes place if the judge suspends the sentence, places the prisoner on parole, expresses confidence and personal interest in him, and thereby instills such resolve that the man is reclaimed and justice itself ultimately better conserved than by the exaction of a pound of flesh."

Justification of man in the face of moral insufficiency is a dimension of religion that passes beyond its role as a handmaiden to ethics, as the doctrinal agent which provides the imperatives and the primary moral values. Here religion fulfills a purpose no other human experience can achieve. Perhaps it is this feeling of frustration and humility before the *ought,* transformed into a joyful sense of forgiveness by unbounded love, that touches the very core of religion. And to complete the story—even though the fact is irrelevant in this connection, let us note that frustration and humility before *existence,* transformed into wonder and awe at the marvels of nature, form the other, parallel avenue to the heart of religion.

This state of separation of ethics and religion, suggested by some of Luther's utterances and described here as one phase of the reformer's views, is a clean philosophic possibility, but it was not his total commitment. For Luther's God was not only the God of grace and mercy, he served also as author of the ethical commands in accordance with the teachings of the Bible which Luther accepts as authority. Indeed the early part of Luther's life stands wholly under the impress of the commanding, the terrible God of wrath, the meaning of the Cross being a later discovery which resolved his early neuroses and inspired his protestations against papal rule.

The effective difference between Paul and Luther is therefore small. Both obtain their imperatives from dogmatic religion. Paul wants universal love to be the medium in which moral actions are suspended, looking to Christ as the exemplar for such love. Luther takes the Commandments more seriously, worries more about their meaning and, in his catechism, applies them in definitely casuistic fashion to the affairs of daily living. In some instances he becomes most specific: "Thou shalt not steal should be placed by the miller on his sack, the baker on his bread, the shoemaker on his last, the tailor on his cloth, and the carpenter on his ax." It is this unusual preoccupation with specificities that reveals to him the full disaster of human moral incompetence and forces him beyond justice, beyond love upon divine compassion.

Here we have clearly left the field of ethics. In the example of Luther we have seen how ethics, more specifically the realization of man's ethical helplessness, can induce religiosity.

ETHICAL SYSTEMS WHICH ARE LOOSELY COUPLED WITH METAPHYSICAL VIEWS

Confucianism and Buddhism are almost pure ethical systems in our sense, and Taoism with its mildly metaphysical Yin and Yang comes close to this description.

Confucius. Confucius[10] was a statesman and savant who shunned all absolutes, whose precepts are sometimes regarded as the forerunners of modern humanism and, by some, as an early form of positivism. His norms impose universal benevolence in the form of practical charity, helpfulness and generally pleasing attitudes toward others; they forbid lying, cheating, stealing, and admonish the Chinese to observe the social etiquette of his class. Confucianism is definitely tailored to the economic situation in which it arose and which it attempted to stabilize, and it makes no

[10] See the *Analects of Confucius,* Transl. by A. Waley, London: Allen and Unwin Ltd., 1938.

claim to transcendental wisdom. In particular, the Chinese have often displayed the good sense to discern that there are situations to which ethics, i.e., Confucian ethics, do not apply. And why are these rules to be obeyed? Confucius' answer is remarkably direct and simple: His rules hold the members of a family together, they have produced satisfaction in the lives of the ancestors. The imperatives are there. Their postulational character is never hidden in metaphysical verbage; they are simply stated as the revered teacher's opinion—almost every paragraph of the Analects begins with the phrase: The master said—and that seemed to be enough. The principle of validation in Confucianism is human happiness on this earth, and the reason for accepting this scheme is that previous generations have proved it successful. Hence the unusually strong emphasis upon tradition as evidence for the rightness of the code.

Here, then, is a very close approach to what has been called in the preceding pages a "pure" ethical system.

Buddhism. Buddhism is the atheistic, essentially ethical termination of a long development of religious systems in India. Its prior phases are the complex polytheism of the Vedas and the profound doctrine of the Upanishads known as Brahmanism, now degraded to a popular kind of polytheism. The Buddha stripped Brahmanism, whose main thesis is the identity of Brahman and Atman, the universal soul and the individual consciousness, of many metaphysical trappings, and activated the theory within concrete experi-

ORIGINS OF IMPERATIVES AND PRIMARY VALUES

ence. He announces rules of behavior, saying in effect that if you try them you will like them. As the goal of ethical actions, as the primary value, he holds out the most blissful state, which behavior in accordance with his imperatives (the eightfold path) is supposed to bring about. The correctness of his claims, their practicality, will not be questioned here because it is the form of his thinking that matters in the present context. Nor do I wish to complicate the issue by noting the controversies which are taking place, even among Hindu scholars, regarding the precise meaning of nirvana. Whatever our thoughts on these questions may be, Buddhist ethics have been amazingly effective, and they display almost in pure culture the characteristics of a system patterned after science.

To be sure, Buddhism has other important tenets which, because they are irrelevant to our inquiry, we are forced to ignore in this extremely crude and brief appraisal. Nirvana is not attained in a single life but only after a series of reincarnations. There is the doctrine of Karma, of rigid fatefulness pervading the sequence of lives, which must be broken through ethical attitudes and actions before nirvana can be attained. These are complications which leave the empirical structure of Buddhist ethics intact.

SUMMARY AND CONCLUSIONS

The present chapter was entitled Historical Origins of Imperatives and Primary Values. We have given a very cursory survey of various ethical systems from that vantage point, restricting our view principally to moral doctrines which have proved their efficacy in history and omitting for the most part the views of individual philosophers who, though of perennial fascination to writers on moral theory, have had but little effect on the development of our race. While the general structure which admits of a distinction between imperatives and principles of validation is present in nearly all important systems, the bewildering variety of origins for both of these is especially noteworthy and indeed amazing.

As to the imperatives, they frequently come into being by the fiat of some authoritative law giver in an historically recorded, seemingly creative act. In other instances, practices developed during centuries are frozen into a formal code by one historic manifesto. Or it may be that the last process is the common one, the former, the sudden generation of norms, being the only part of an historic process whose record is preserved. Most impressive here is the role played by forceful individuals.

The principles of validation, on the other hand, are in general less clearly formulated; they are rarely associated with specific names, although they can always be seen in action. Men have at all times had answers to the why of their ethical behavior, but there has been greater apparent diversity among these answers than among the accepted commands: Thou shalt not lie, steal or kill are nearly universal laws; systematic knowledge of the importance of principles of validation, and indeed an active concern for them, seem largely lacking.

This is perhaps not difficult to understand. In a world separated into nearly isolated cultures the paramount moral problem is the firm establishment of those *de facto* values to which history has committed each culture more or less by chance. This can usually be accomplished by reliance on tradition, by citation of ethical examples, by an emphatic recital of appropriate commands together with vague promises or threats. References to validating procedures are to a considerable extent unnecessary and therefore the procedures themselves are rudimentary.

But when contacts between cultures, between different systems of values, increase by reason of higher social mobility, rapid and easy communication, then the supercultural question: which system of values is best, inevitably arises, and when ideological allegiances are strong, it clamors for an answer with dreadful urgency. It is then that empirical criteria of the kind we have suggested must enter the scene, that insubstantial logic and mere preference cease to suffice as bases for judgment. Hence it is our present era which demands, more than any other, that we become clear about our criteria of acceptability of ethical norms and, having once formulated them, that we exhibit some consistency in applying them.

When this is realized the enormous magnitude of the problem confronting us is exposed. For it will then become evident that mere choice of a principle of validation (or a compatible set of them)—and agreement on them is probably not difficult to attain—is not sufficient because the terms in which the principles are couched are intolerably vague. Happiness, liberty and even peace mean different things in different cultures. Hence the first step in the approach to universal ethics is a semantic one, namely the clarification of such terms and, if we follow once more the lead of science, their *operational* definition. Strange though it may seem, it may be that before we can reach any political understanding with our communist adversaries we must begin by discussing with them the empirical meaning of human happiness. This lengthens the road to the attainment of supercultural ethics, the road to peace; but I doubt if any short cuts are safe or even possible.

Turning from prospect to retrospect, I wish now to place into view the similarities between the historical development of ethics and of science. In science, too, the basic postulates were usually conceived by the genius of a single individual. These, we recall, correspond to the imperatives of ethics. Here also conjectures which evolved gradually in scientific thinking often crystallized into formal theories on definite occasions. Among the documented instances of this sort are the conception of Pythagoras' theorem after the use of the 3–4–5 rule of the Egyptians rope stretchers who surveyed the lands along the Nile; Newton's formulation of the law of universal gravitation after the gropings of Descartes and

Kepler for an inverse square law; Helmholtz's postulation of the energy conservation principle following a period of condensing doubts regarding the possibility of perpetual motion; Kelvin's conception of the second law of thermodynamics in response to the frustration felt among scientists who failed to understand the meaning of irreversible phenomena. These are minor incidents, perhaps, in comparison with the conception of ethical systems, but the circumstances of their occurrence are similar.

As for scientific attention to principles of verification, which are the counterparts of primary values, we find that it has also evolved quite slowly, being retarded by lack of evident need. Prediction was generally accepted as the purpose of science, but precisely what was meant by prediction, the conditions under which that purpose could be said to be achieved, are matters cleared up only through the latest developments in stochastic theory, and even there only partially and with postulational acceptance of certain propositions concerning errors (See Chapter I). The overall parallelism is again quite striking. There is no general doctrine of "theory formation" in science, just as there is no general *a priori* prescription for the establishment of a workable ethical system. Both owe their origin for the most part to creative intuition, which defies prediction.

VIII

Consequences of the Parallelism Between Ethics and Science

ABSTRACT

This chapter takes a second look at values and locates them in the middle range between imperatives and principles of validation. It recognizes ethical conflicts as resulting for the most part from ambiguities of postulates, reveals conscience as the instinctive residue in man's mind of the results of the ethical experimentation of the race through history. Various types of ethical relativity are examined and traced to their methodological roots, and the question of obligation vs. purpose is resolved by noting that these two terms designate complementary aspects of moral suasion, not a clean alternative. The last section contains comments on the actual substance of the modern ethical code.

A Second Look at Values

The present view confers a fresh meaning and a new vitality upon values as it allows them to be seen in a large methodological perspective. The résumé in Chapter III has already led us to suspect that values are not given per se but arise as byproducts within the ethical enterprise. To be sure, our interest is here limited to moral values, or virtues, to qualities like goodness (in a certain specific sense; see below), honesty, trustworthiness, kindliness, altruism, and justice as well as others that go beyond the idea of virtue, such as human life, health, wealth, and friendship. Ideas like beauty and truth are not included, for they refer to circumstances and involve experiences wholly incomparable to those of moral values, and any theory which confuses them with virtues or attempts to establish them on an equal footing with moral values must fail because of a serious oversight at the very start.

For consider beauty, truth and honesty. They are commonly called values. But what they have in common is merely their ability to be present in greater or lesser amount, their susceptibility to being graded. One painting can be more beautiful, one statement more true, one person more honest than another. If this characteristic of gradability is to define values that term is practically wasted, for it would then have to be applied to numbers, to all measurable quantities, to all emotions. The restricted theory of values in vogue among most philosophers will thereby lose its point.

I suggest that the differences between beauty, truth and honesty are far more significant than their trivial common trait: beauty is primarily a property of objects (although it is also ascribed to pleasant experiences), truth adheres to propositions, honesty reveals itself in human actions. The immediate unreflective language of beauty is the ejaculation, that of truth is the indicative, that of honesty the imperative, and I take this to be significant even in view of the fact that language is not itself primary but only symptomatic of the character of underlying experiences, even while granting that every occurrence of beauty, truth and honesty can be expressed in indicative sentences. The subjective fabric of beauty is, nevertheless, emotion; that of truth, intellection; that of honesty, obligation. It should, therefore, be clear at the outset that there can be no purely emotive theory of truth or of ethics, no purely rational theory of beauty or of ethics, and no hortatory theory of beauty or of truth.

We therefore treat in this context only the homogeneous class of attributes called *moral values*.

Here we encounter first of all the primary values: happiness, benevolence, universal love, perhaps survival. These, as we have seen, are postulational or arbitrary—if that word can be deprived of its derogatory flavor. We do not possess a rigorous or even a full descriptive theory of primary values; certainly the pages of this book can not pretend to furnish one. Consequently we do not know except through vague intuition whether, for example, the primary values just listed are indeed independent, or how they are related in detail. That happiness presupposes survival is fairly clear, but whether happiness occasions benevolence or the reverse, or whether these are intrinsically unrelated, can hardly be settled with finality. In further pursuit of ethical researches it would seem mandatory that such problems be illuminated, not only by logical but also by empirical studies.

Primary values act in our methodology as principles of validation, deciding the rightness or wrongness of the initial commands, furnishing the ought or the veto which in turn establish obligations. Between these two poles stand other values, either presumptive or confirmed, like milestones along the road of ethical progress. For if your code contains the imperatives, "Thou shalt not steal" and "Thou shalt not bear false witness," *honesty becomes a value.* As such, and in what I have called its presumptive form, it is a mere logical consequence of the commands; conversely, honesty would not in general be a presumptive value if the code enjoined its adherents to lie to their enemies. In our case, honesty becomes a *confirmed* value if it realizes the primary value, e.g. collective happiness, which has been chosen.

The commandment, "Thou shalt not steal," when taken alone engenders the value called property or wealth; "Thou shalt not kill" makes human life a value, and these values along with innumerable similar ones *take on obligatory power if their realization in human actions satisfies the principle of validation*. The reader may find it an interesting exercise to trace the values listed in the first paragraph of this section, trustworthiness, kindliness, altruism, justice and friendship to particular imperatives in Western moral codes.

The epistemology of values bears a strong resemblance to that of facts which are often taken to be the counterparts of values in science. Neither is primary, unanalyzable or self-generating. Facts arise during the course of cognitive procedures between the postulates, i.e. the general principles of science, and verification on the P-plane. In saying this I am actually using the word fact in a specific manner, rescuing it from indiscriminacy just as the concept of value had to be rescued: Thus, we speak of the action of gravity as a fact in the physical world, saving the word to mark a certain universal aspect of nature in contrast to the trivial observation: this stone falls. Now, the force of gravity, which we called a fact, is not anything directly perceived, nor is it written in capital letters in the sky, nor is it a Platonic form. It is something which derives from Newton's universe-square law and whose consequences, via rules of correspondence, are in turn successfully tested in observations. The distinction between presumptive facts, facts merely entailed by postulates, and confirmed facts or verifacts, which are both entailed and verified, is also appropriate in this connection.

The word "good" is often regarded as denoting the supreme value, moral and otherwise. Discussions concerning it have, in my opinion, descended to the lowest intellectual levels in philosophical debate. No further reference will here be made to the strange dispute as to whether goodness is a natural or a non-natural quality, for the simple reason that preoccupation with natural science has failed to enlighten me with respect to the meaning of "natural," or to the distinction in question. Nor can I see any one single meaning in the word good. Rather, there seem to be two large complexes of ideas, each more or less coherent, to which that word commonly applies. First, it designates something either material or abstract which one likes. This meaning of good is emotive, entails no obligation and can be treated fully by the so-called emotive theories of ethics. But it is uninteresting in any truly ethical context.

Secondly, good describes human actions. Here it does take on moral significance and moral value. For it means *right,* it means to be in accord with the imperatives of an ethical system known—or at least expected in the fact of overwhelming evidence—to satisfy its principle of validation. A confusion between these two meanings of good accounts for the erroneous claim of the emotivists, who are able to deal with the first aspect of good but not the second, to be in possession of the clue to moral values. Inasmuch as the present definition of the moral good reduces it to the notion of right (Greek: *deon*) our view may be called deontological, to use a current terminology. But we do not hold with Sir David Ross,[1] "that an act ... is ... right, is self-evident." It is right if it fits into a scheme such as was outlined in Chapter IV.

Summarizing, then, there are two kinds of moral values: a) those stipulated by commitment to the principle of validation, which were called primary values; b) the secondary values ordinarily called moral values, which emerge in consequence of the imperatives and take on their ought by validation of the total system. Whether a value is primary or secondary may depend on the initial choice. For example, life is a primary value if the principle of validation is merely survival; it is secondary in the case of hedonism. Yet certain values, like health, wealth, kindliness, and honesty are almost always secondary because their scope is hardly large enough to serve as principles of validation.

ETHICAL CONFLICT

Ethical conflict is of great interest to the moral philosopher because it brings into focus many traditional problems in his field. For under conditions of conflict, when mores and ordinary precepts fail, basic principles must come into play, and it is assumed that the nature of these principles can be recognized if they are observed in such instances.

Viewed with methodological detachment this makes ob-

[1] D. Ross, *The Right and the Good*, Oxford: Clarendon Press, 1930.

vious sense. Careful reflection, however, reveals a different picture. It shows first that the frequency of conflicts in ordinary life is low and that their importance in ethics is likely to be overrated. What parades heroically as conflict is often a clash of interests in which the voice of ethical imperatives is clear but unpleasant, or else it is a command that is imperfectly understood or lacks casuistic explication. Conflicts are second-order effects in ethics, to which attention needs to be drawn when first-order matters, such as clarity and explicitness of statement of the ethical rules as well as persuasion to secure commitment, have been accomplished. Their seeming importance is doubtless related to our fascination by tragedy, which features ethical conflicts to a degree unusual in life.

With this perspective let us analyze a few typical conflicts. We consider three situations, none of them of very common occurrence as ordinary ethical problems go, and then see how their difficulties can be resolved.

1. A madman tries to kill a defenseless person. It is clear beyond conflict that I should prevent this act. But suppose I can prevent it only by killing the madman. Should I kill him to save the other person?

The answer is *no* if I take the commandment, "Thou shalt not kill," literally. It is *yes* if that commandment is subjected to the qualifications by which our society actually restricts its application. There would be no conflict if the fifth commandment were properly formulated in my mind, if I had been taught the circumstances under which the imperative is to be suspended, or if I knew that there is no qualification. One of the defects of our customary imperatives which the present approach exposes (and to which earlier allusions have already been made) is their bluntness, their unintended appearance of universality, their failure to be casuistically qualified. Here again, we may well learn and take comfort from science. Its postulates, the counterparts of our imperatives, do not pretend to universal applicability. Every textbook which teaches them is careful to state the accessory conditions under which they hold. The principle of universal gravitation is effectively suspended in favor of Coulomb's law when we calculate the force between two electrons. We regard this as normal and do not speak of a conflict between scientific principles. Why should the situation be different in ethics?

2. There is the problem of the conscientious objector in time of war. If he takes the fifth commandment literally, without qualifications, he may not join the army, and there is no ethical conflict. Conflict nevertheless arises between his ethical decision and the law of the land, which is unethical unless the fifth commandment is suitably qualified. Again, there is little to be said from a fundamental ethical point of view, except to note once more the paramount need to clarify the meaning of our imperatives.

3. A man feels forced to steal to keep his family from starving. Here again is an ethical conflict in which the

obligation to support one's family opposes the imperative, "Thou shalt not steal." If he takes it literally and without qualifications his course is clear; his family starves. With the qualifications tacitly accepted and approved by our society the opposite course is equally clear.

It is seen that conflict arises in all three instances from incompletely stated and unqualified imperatives. Now there exists a widespread feeling that moral imperatives must be simple, that qualification detracts from their dignity and makes them into laws. It is difficult to find rational arguments in favor of this sentiment. Law is, after all, applied ethics, and greater effectiveness is not tantamount to loss of dignity. True, we prefer simple statements, in science as well as in ethics, and we have seen in Chapter I that simplicity is one of the guiding principles in every search for scientific explanation. But simplicity must be subjected to the procedures of verification and in that process it is often partly sacrificed. So in ethics: imperatives must be as simple as possible *as is consistent with validation*. Here, too, "simplicity should be desired but also distrusted."

To some, the present suggestion for the disposal of conflicts will seem to beg the ultimate value question; they will ask: does not the resolution of the three conflicts above, effected by casuistry (the explication of imperatives) presuppose a hierarchy of values? In example 1, do I not judge that the life of the defenseless person has *greater value* than that of the madman? In 2, does not the usual solution involve the belief that the life of my fellow countryman is more valuable than the life of an enemy? In 3, do I not affirm that life has greater value than property?

These judgments are correct, but we have stated what happens in them incompletely. Our reference to values is not a priori, independent of the other elements of our ethical system. When I say, the life of a normal person is more valuable than the life of a madman an informed glance is being taken at the principle of validation, which is evidently some form of collective happiness, and the value judgment reflects elliptically a conviction instinctively engendered by that glance and along with it all the suggestions of history, literature, and conscience which have set the stage for that conviction.

A similar analysis holds for the other two examples.

Conscience

As an active monitor, the human conscience is more powerful and a greater influence toward moral good than all ethical theories that have been proposed. Interest in it, however, seems to have been lost by most modern writers in the area of moral philosophy and one wonders whether this circumstance may be related to the increasing sterility of moral literature. It seems that an account of ethics is incomplete unless it attempts an explanation of conscience; hence we ask: How is conscience related to the principles outlined in these pages? Is it an innate, primary arbiter of ethical rightness, a Kantian practical reason which must be taken as given in human nature? Is it the voice of fear warning us against punishment? Or is it a reflection of the divine in the human soul?

Butler[2] speaks of conscience as "the moral faculty," innate in man, which enables him to formulate true moral judgments. He attributes to it two special functions, one cognitive and the other authoritative, and in this convenient way he bridges the factual with the ought. In its cognitive function, conscience "pronounces determinately some actions to be in themselves evil, wrong, unjust." It manages this with a modicum of ordinary reason—although Butler occasionally identifies conscience with reason—intuitively rather than by lengthy reflection. "In all common ordinary cases we see intuitively at first view what is our duty. This is the ground of the observation, that the first thought is often the best. In these cases doubt and deliberation is itself dishonesty. That which is called considering what is our duty in a particular case, is very often nothing but endeavoring to explain it away."

This, it seems, is too easy a solution of the problem of conscience, indeed an unsatisfactory one reducing to fiat even if it could be accepted on factual grounds. But that is not

[2] J. Butler, "Sermon II—Upon Human Nature," *Works,* London: Macmillan, 1900.

the case, for as Westermarck, the proponent of ethical relativity and many others have seen: different consciences often issue conflicting judgments and conflicting orders. How is this to be reconciled with a universal authoritative moral faculty?

Let us briefly review other theories. Kant's version, as might be expected, is complicated. He calls conscience an inner forum, a last instance of appeal, in which man sits in judgment upon himself. Stoker[3] paraphrases him as follows: "Whether an action is at all (überhaupt) right or wrong, this is judged by reason (Verstand), not by conscience. Conscience does not adjudicate the action as a case under law—this is done by practical reason." In conscience "reason examines itself to determine whether it has truly performed the judgment of an action with all possible deliberation, posing man as witness for or against himself to ascertain whether this has happened."

Here, then, we have a view similar to Butler's and subject to the same objections, but refined and elaborated by the distinction between practical reason and conscience, the former pronouncing judgments on human actions while the latter is a sort of supreme court of errors in which the legality of reason is finally appraised. But in the face of Kant's elaborate theory even the Kantian must wonder why it is that the verdicts of practical reason and of conscience are

[3] R. Stoker, *Das Gewissen,* Bonn: F. Cohen, 1925.

so variable and fallible, while those of pure reason enjoy universal assent.

Schopenhauer's theory of conscience is less exalted[4] and reflects an attitude which is not uncommon today. For him, conscience is knowledge of *ourselves* which through our *own* actions and reflections becomes more and more intimate and complete; it is "das immermehr sich füllende Protokoll der Taten." Objectionable here, it seems to me, is the subjective aspect of this personal register of deeds and reflections; what stands against it is the psychological fact that conscience in children is often a stronger force than it is in adults, hence it can hardly be a growing—and presumably increasingly authoritative—protocol of deeds. Somehow, one feels, Schopenhauer misses the intersubjective relevance of conscience. This leads him to deprecate it, as in his famous passage: "Many a person would be astonished if he saw what actually composes his conscience, which to him seems quite respectable (stattlich): approximately ⅕ fear of people, ⅕ deisidemonia, ⅕ prejudice, ⅕ vanity and ⅕ habit; so that he is basically no better than the Englishman who said, 'I can not afford to keep a conscience.'"

Perhaps the lowest estimates of human conscience were rendered by Nietzsche and Freud, for whom a bad conscience was a disease. Having acquired a strong reluctance to translating the highly individualistic language of these

---

[4] For a good discussion, see Stoker, *op. cit.*

authors I will quote them raw (with assurance to the reader that he will miss nothing of consequence to the thesis of this book if he disregards the quotations). Nietzsche says:

"Das schlechte Gewissen ist die tiefe Erkrankung, welcher der Mensch unter dem Druck jener gründlichsten aller Veränderungen verfallen musste, die er überhaupt erlebt hat —jener Veränderung, als er sich endgültig in den Bann der Gesellschaft und des Friedens eingeschlossen fand." Bad conscience, according to this author is a characteristic of the weak; the "Herrenmensch" is not affected by it.

Freud[5] proposes the following explanation. The child suffers from an Oedipus complex. The superego resolves the complex, takes on the character of the father and thus assumes the role of conscience. This is about as good an explanation of conscience as the thunderbolt theory is of lightning. For further evidence on its utility, note this:

"Was sich hinter der Angst des Ich vor dem Über—Ich, der *Gewissensangst* verbirgt, lässt sich sagen: vom höheren Wesen, welches zum Ich—Ideal (Über—Ich) wurde, drohte einst die Kastration, und diese Kastrationsangst ist wahrscheinlich der Kern, um den sich die spätere Gewissensangst ablagert; sie ist es, die sich als Gewissensangst fortsetzt."

Ascending to the level of more responsible estimates of the essence and the role of conscience, we note first, in the words of Brand Blanshard[6] who follows the view of Leslie

---

[5] S. Freud, "Das Ich And Das Es"; *Massenpsychology und Ich—Analyse*, VII. Vienna: International Psychoanalytic Publishers, 1923.

[6] B. Blanshard, *Reason and Goodness, loc. cit.*

Stevens, that "conscience is the concentrated experience of the race. It makes us wiser than we know, because it is the deposit of parental example, of the instruction of teachers, and of the pressure of society, themselves in turn the product of experimentation. Conscience is thus the voice of our own hitherto accepted ideal, recording its yes or no to a proposed line of conduct. It does not in general argue; it simply affixes its seal or enters its protest."

To this, only two footnotes need to be added to make the story complete. One has to do with the character of the concentrated experience of the race, in particular with the way in which it takes on normative qualities. I believe that this experience, of which the conscience is the residual recorder and monitor, is cast within the framework of avowed commands and principles of validation. Without these the experience of the race would be amorphous, ethically indiscriminate and lacking normative force. But if they are present, albeit in vaguely formulated intuitive conceptions of divine imperatives and ideal goals, then good and bad experiences can be distinguished and record themselves. Conscience, therefore, is the residue, itself both memory and prompter, within an individual mind, of mankind's (or a limited society's) past experience in ethical achievements and frustrations, a record of its slow movement from imperatives to goals.

The second footnote enlarges the concept of the "deposit" of which Blanshard speaks. He modestly supposes that it occurs as a result of the normal processes of teaching and communication. It seems that it must be more than that, more by virtue of a mysterious transmission of abilities and unformulated knowledge which, although as yet unfathomed by science, is more effective than ordinary communication. Conscience is a deposit of the past experience of the race in the same sense in which the ability and the disposition to fly is a deposit, in a newborn bird, of the experience of flight among birds. To admit, as one must, that this makes conscience an *instinctive* moral guide merely acknowledges our ignorance of the scientific genesis of conscience, but leaves its self-declarative authority intact.

This view clearly makes allowance for that measure of relativity in its judgments upon which Westermarck insists. Different cultures have slightly different codes and primary values; individuals differ in their endowment with instincts, and very largely with respect to the intensity of assertion. In terms of these observations, the ineffectiveness and the fallacy of conscience in individual instances can be explained. And its instinctive nature accounts for the strength of its voice in children and for its tendency to become fainter as men begin to reason.

Ethics as a Group Enterprise

The method of ethics may conceivably be applied to a single life, and it often is. An individual may decide that his goal is personal pleasure and he may embrace certain rules of conduct which he deems conducive to it. Or he may desire wealth and choose to acquire it by any means including stealing. Suppose he succeeds; does this mean his ethical system is validated?

Here again science, the methodological forerunner of ethics, provides an unambiguous answer through analogy. Not, to be sure, the science of the last century, the cocksure dispenser of ultimate truth. The answer lies in one of the most recent lessons learned by science, learned last by physics which prided itself on rendering the most perfect description of nature. The lesson is this: a single occurrence, a single observation does not confirm a law.

The lesson was learned in two stages. The first accompanied the development of the theory of errors. Suppose a scientific hypothesis requires that under specified conditions a certain substance shall reach a certain temperature, say $135.2°F$. Observation reveals that the temperature is $175.0°F$. The experimenter becomes dismayed, wonders whether something went wrong in his handling of the apparatus, whether he mistook a reading or blundered in putting his chemicals together. At any rate, he does not give up in his attempt to "prove the law" he conceived, for he would love to demonstrate it to the world and enjoy his success quite as much as an artist enjoys acclaim. So he repeats the experiment with greater care, obtaining this time a temperature reading of $173.5°F$. His spirit sinks, but he makes one further attempt and gets $176°F$. At this point he abandons his efforts and his hypothesis and returns, greatly saddened, to other tasks. A *single* outcome had not convinced him of the error of his ways.

But consider this happier sequence of events. The chemist, expecting a temperature of 135.2°F, reads 136.1°F. His spirits soar, he is greatly encouraged but not convinced. This and nothing more would also be the case if he had read exactly 135.2°F. On repeating the experiment he finds a temperature of 134.7°F. As he continues—and he must continue to make his case—he attains values all of which lie in the neighborhood of the expected one, even though none coincides with it exactly. In the end, however, and after performing some statistical calculations, he announces that his law is true and takes a creator's joy in his discovery.

Why was the law not exactly verified in every instance of observations? Because man is prone to errors. He never puts together the precise amount of chemicals required, he cannot read a balance with perfect accuracy, indeed he cannot build a perfect balance; even his thermometers are not exact. Because of all these inevitable contingencies the final result must be in error. There is more to this story than appears here. The mathematician can actually calculate the way in which the final results should be distributed if their variation is due to numerous small errors, and that distribution can be checked and has been checked successfully in many cases. The conclusion, then, is clear. There *exists* a true value of the measured quantity, and in the present instance it would be a temperature of 135.2°F. But man's limitations and the crudeness of his apparatus prevent him from observing it except occasionally by accident.

A single instance does not confirm a law. True, we say; but the departures from the value necessary for confirmation are so small and the cause for the departures is so trivial that this lesson is insignificant for ethics. This comment is likewise true, but it characterizes only the first stage in the current stochastic involvement of science.

The second stage was initiated by Heisenberg's discovery of the uncertainty principle. Its practical importance is greatest for the most basic and elementary processes which take place in the atomic and nuclear realm, and its correctness has been amply demonstrated. It says in effect that departures from lawful values of physical variables can *in principle* not be avoided, that every observation, no matter how carefully contrived, will manifest an unpredictable variance from expectation. While the physicist, beholden to his past, still speaks of errors even in connection with the uncertainty principle, he knows full well that the word has lost its usual meaning, that nature itself and not the experimenter is responsible for the vagaries of elementary observations. Nor are the departures from the mean necessarily small in atomic measurements. A particle of known momentum will exhibit in different observations designed to locate it in space, infinitely large variations in its position, none of which could be predicted. Physical law nevertheless has something positive to say about its position: it may for example specify the arithmetic mean of many measurements. To verify that law, the scientist cannot content himself with one or two or even a dozen observations: he must make millions, as he does when he determines the impacts of a swarm of elementary particles, e.g., photons, on a photographic plate.

Verification in science, on the most basic and elementary plane, is intrinsically a stochastic process, a multiple observation, a mass phenomenon. This is the hint we were seeking, and we now apply it to ethics.

Validation in ethics, we hold, is likewise a mass phenomenon requiring a group of actors, and the larger the group the better. Individual ethical experimentation is as meaningless as an observation of the position of an elementary particle. The man who steals and gets away with it proves nothing.

Earlier we criticised Kant's categorical imperative, showing that it is not an imperative at all, nor even, standing alone, a principle of validation. We now recognize it, however, as an appreciable contribution to ethical theory inasmuch as it places group validation in the focus of ethical interest. Only if an action is performed by everybody—or nearly everybody—in a society and leads to the primary values does it take on normative significance. Individual acts of heroism are admirable, but they may prove nothing of ethical importance.

There is a converse to the conclusion we have reached.

For if ethical validation is indifferent to isolated behavior, isolated misbehavior must also escape the reach of ethics. That is to say, behavior not embedded in a group is ethically neutral. Perhaps this conclusion is academic, but it should be stated for consistency. What it means is this. If a person were stranded on a remote unpopulated island where he is sure to die before he could possibly make contact with other men, ethics would lose hold upon his actions. Other agencies would control them; he might act from religious motives, or become despondent, or seek a painless death.

Admittedly, the analogy we have drawn with science does not *prove* that ethics is necessarily a group enterprise. The evidence that it is comes from anthropology and from history. Still the analogy permitted us to make the issue more precise, and its very existence exposes a deeper unity in human nature: man's quest for knowledge and man's control of action receive their sanction in similar ways, one might say in social ways, if the multiplicity of scientific objects may be called a society.

If one rules out individual ethical experimentation, does one not thereby inhibit or proscribe ethical leadership? That this is not the case will be seen in the next section.

Ethical Relativity

In the moral field the word relativity is often used with a meaning entirely different from that employed by physicists. For it denotes permissiveness, the right of a person to select his principles of behavior in accordance with his special circumstances, or in a milder sense the evident cultural dependence of moral norms. In physics, however, relativity means the *invariance* of nature's basic laws with the consequence that *special* observations, like the recording of distances, time intervals and speeds are conditioned by the system of reference in which they are made. Physical relativity does *not* entail that one person's laws are different from those of another, that an observer in motion sees the world differently than an observer at rest; quite the contrary: relativity denies that motion has an effect on the nature of things and it makes the appropriate provisions in stating its laws to make that denial true.[7]

Occasionally one finds a professional moral philosopher reproaching Einstein and others for having released the virus of relativity which has infected, not only science but even ethics, and has eaten away the marrow of its authority. Protagoras, one hears, was the first relativist because he said "man is the measure of all things," Einstein merely invented the mathematics and refined the theory. Nothing could be further from the truth, as a perusal of Frank's book will show.

Now it happens to be true that ethics are relative, or relativistic, in a way in which the laws of physics are not. Mores do differ from place to place, and in a given place

---

[7] See P. Frank, *Relativity, A Richer Truth* or H. Margenau, *Open Vistas*, Chapter IV.

they differ from time to time. So do the imperatives, though their variations seem more restricted, and the principles of validation. This variability—a word which probably expresses the ethical state of affairs better than relativity—is to be expected if the view proposed here is correct. For it may occur as a result of one or several of three causes: a) incomplete understanding of the imperatives and the primary values; b) changes in the commands made as time goes on in an effort to improve the adequacy of the ethical system; c) different choices of postulates in different cultures.

Of these, a) induces variability between individuals within a group. This is clearly a situation short of ideal and in need of correction when it occurs. The remedy is what was called moral teaching and properly conducted casuistry, an explication of the norms in terms of specific instances and insistence on their being observed. This kind of variability, while ever present, is objectionable.

The changes resulting from b), however, are normal. Every ethical system, conceived in the constructivist terms of the present essay, is necessarily capable and in need of improvement. Like truth, it is posed as a problem which is being solved with greater and greater refinement. Here is the place for the moral leader, the lawgiver, the martyr who is willing to die to assure men of his conviction that their imperatives are evil, a St. Francis or a Schweitzer who wish to extend their ethical obligations beyond mankind to all life.

Let us repeat; it is natural and wholesome that in the course of time the normative system of any given culture undergo some changes. A stagnant set of external values indicates nothing but lack of vitality and is to be deplored. If this progressive change is ethical relativity, let us not be without it.

Finally, as we come to cause c) we note that behavior can differ intrinsically among people because of different choices of imperatives and primary values. There is no obvious reason to suppose that several of these, which differ to the point of contradiction, may not be validated in human living. If this is true there are several sets of "oughts" between which there can be no reconciliation. We are thus face to face with a perplexing, an embarrassing possibility which seems to bring our whole approach to naught.

In a moment I hope to show that this eventuality, while sobering, is not as destructive as it might first seem. Meanwhile it is important to realize that its occurrence cannot be recognized until the premises of two opposing systems are clearly formulated and fairly understood. One is all too prone to suppose that the ethical system associated with communism is so radically different from that of our Western democracies that the debacle is upon us and one feels there is no hope for the development of a common pattern of behavior. This judgment is a hasty one, for it ignores the present no-man's land which lies between imperatives and principles of validation. Are we clear in our own minds as to what our imperatives are and what they mean with respect to daily living? Do we agree on our primary values? Our system may appear to differ from others because we, or they, have not thought through the consequences of our or their declared tenets. The effort to do this will bring clarity and add confidence to our own beliefs, insure deeper understanding of theirs, and it may indeed indicate that the basic tenets of the two systems are not nearly so far apart as we conveniently but thoughtlessly supposed. There is, in fact, every evidence that the primary values of the Russians are the same as our own, and that they share our ethical commands. Differences, sharp differences occur in the area of secondary values, where words like freedom, subversion, patriotism, loyalty are bandied about in artful but illogical fashion. What is needed above all is an illumination of the consequences of our basic commitments, and this may show that the customary easy declaration of incompatibility among ethical codes is a falsehood born of ignorance with respect to the meaning of our ethical commands.

This is not an ill-conceived Western speculation. Eastern scientists are passionately interested in the basis of human morality, and I have had the pleasure of learning on more than one occasion that they discuss the methodological foundations of ethics with eagerness, objectivity and profundity. Many of them hold that it is idle to attempt agreement on political matters before these basic issues are worked out, that political agreements are bound to be pacts of prudence, expediency and opportunism until the ethical foundations are clear.

We see, therefore, that real conflict between systems can hardly be discovered until the fog of casuistic misunderstandings is dispersed. But now suppose this to have occurred, and furthermore assume the systems to be truly at odds. I imagine this took place when Christianity impinged upon the heathen world, when Hitler exploded over Europe. In the latter instance history has settled the question of the adequacy of Nazi imperatives to primary values. For while survival is not a highly sophisticated principle of validation it is an effective one and has ruled the Nazi system out. In the case of Christianity vs. paganism the outcome is not quite so assured but indications are hardly lacking.

If political survival were the only or the major criterion of validity in ethics our method of validation could be paraphrased as "might is right." This is not being proposed here for an instant. Ethical and happy people have been destroyed by brutal conquerors, and yet, in the common appraisal of humanity their fate was not an ethical test: men will again and again live by the precepts of the saintly nation which failed to survive because they acknowledge higher criteria than survival.

It is difficult, also, to think of an historical example illustrating the symbiosis of contradictory and valid ethical systems. Christianity and Buddhism will not serve, for while they differ radically in their religious conceptions, their ethical bases do not differ at all. The case that gave us pause is distinctly academic.

But it can occur. There is then no further opportunity for arbitration; we are saddled in that case with two conflicting sets of oughts. Ethics is an ambiguous enterprise, not by virtue of the necessary choice of commitments for validation, but because of indiscriminacy in the process of validation itself. The ethical universe is double-valued and we can live in either part of it, though not in both.

This is the sort of thing which science faces continually, albeit in limited domains. Rival theories spring up at the forefront of research, where known facts are insufficient to decide between them. There is indeed no a priori assurance that such scientific ambiguity may not persist forever! Conceivably two different models may be equally competent to explain all known facts. When this happens the scientist has no alternative but to wait until further evidence enables discrimination. In the past, scientific ambiguity has never existed very long, and the scientist has now acquired a belief that the progress of understanding *is* unique. In ethics, the scientific outlook is moderately new and unfamiliar, the process of validation incomparably slow; hence confidence in the uniqueness of the ethical enterprise has not had a chance to grow, and perhaps the search for unattainable eternal values has retarded it.

In sum: ethical relativity is variability of norms. Such variability may occur in time within a given culture; this is proper and indicative of the groping for deepened ethical understanding. It may occur "in space," that is to say from individual to individual within a given culture; this is an intolerable state of affairs, occasioned by defective understanding of the ethical postulates or by halfhearted commitment. Variability between cultures is normal but must in time be overcome. At the present stage the major remedy seems to be clarification of the nature and implications of each culture's norms. It is folly to ascribe such cultural differences to conflicting postulates before their meaning is made manifest.

Obligation vs. Purpose

The question whether the good man acts from duty or in expectation of some sort of reward is often asked in the belief that it poses a strict alternative, that the answer must be one or the other. If our description of the ethical task is correct the alternative is fictitious; obligation in the form of abstract duty and attraction by a goal both play their parts. Obligation is an aspect associated with the imperatives, purpose enters through our commitment and our hope to attain the primary values.

The reason why tradition demands a choice between these alternatives lies, I believe, in its failure to see their complementary function. For it has either supposed, as we have shown in Chapter VI, that one of them is self-sufficient and the other superfluous, or it has taken one to be entailed by the other. Having recognized that imperatives are not logically implied by the principles of validation, nor the converse, we are forced and entitled to appeal to both duty and purpose in motivating ethical behavior.

Living systems of ethics reflect this double appeal. Listen to Lao-tse's plea: Be humble, and you will remain yourself; or to the Buddha: abstain from desire and you will earn nirvana; or to Jesus: Blessed are the meek for they will inherit the earth. A command, expressing duty, is always coupled with a cryptic but striking purpose which it is said to achieve.

Throughout these pages the emphasis on the parallelism between ethics and science has been heavy, and the time has come to comment on an important difference. Clearly, the statements above do not resemble sentences of science, which seem to require no reference to purposes. We do not say: apply the law of inverse squares and you will make a correct prediction of the intensity produced by a light source—the last part of the sentence is implicit, unnecessary. The reason for the difference lies in the fact that ethics needs to plead, persuade, cajole, praise, reprove, and condemn, whereas science states and rests its case on its own evidence.

Such disparity has its origin in the contrast between the *est*-laws of science and the *esto*-laws of ethics, and in related differences stemming from this contrast. Both in science and in ethics one meets discordances between the laws and the behavior they regulate. In science one calls them errors, in ethics immoral acts. The measure of the competence of the scientific law to create compliance is its exactness, for the ethical law it is its efficacy. Science seeks to enhance the exactness of its laws, ethics their efficacy. But these are wholly different tasks.

In natural science, what is given and to be subsumed under laws is fixed and incapable of adjustment. Hence, to produce concordance with a law, the law must be altered to suit the phenomena, and the law suffers alteration without pleading. In ethics what is fixed, or relatively fixed, is the set of postulates; concordance is achievable only by altering the actions of human beings, and human beings require persuasion.

What means ethics uses in practicing moral suasion is a question of importance in the study of the history and enforcement of moral teachings. We cannot deal with it here. Suffice it to say that, like any important task, moral suasion often avails itself of aid rendered by outside agencies. We thus find ethics seeking to improve its efficacy sometimes by using religious motivation, sometimes by authoritarian coercion, often by an appeal to punishment, and occasionally by certain sentimental, humanistic rationalizations. These means are peripheral to ethics, just as pedagogical devices are peripheral to the subject matter which is taught.

Conclusion

Questions of method have been in the foreground of our discussion throughout this book, and the reader may feel that the substance of ethics, the actual imperatives and the primary values, merit more attention than they received. This feeling is surely justified, and I make amends before closing. However, lest more be expected than can be delivered, let us recall once again that the valid substance of ethics can not be perceived at the outset; since validation is an historical process, not yet completed, we can only surmise what the composition of that substance in view of present evidence may be.

Among the questions which are left open is the major one: Can there be a multiplicity of valid ethical systems? As we have already seen, there is no logical or empirical reason for rejecting this possibility. If in the course of events it is realized, and if the valid systems are truly contradictory, moral philosophy will have reached an impasse. A similar thing could happen in science, although it never has.

Instead, there is evidence that the laws of science converge, and the belief is general that in the end, if that end ever arrives, there will be one all inclusive scientific explanatory theory. Are we not entitled to the same kind of extravagance in our hope for ethics?

Such hope springs from a rational consideration. The method of science as well as that of ethics is rooted deep in human nature, and human nature has common traits which unite nations and cultures. The method of science reflects the principles of reason which express themselves in the laws of thought. This same reason, channelled in other ways, and allowed to run its course, will develop laws of action. These laws are the product, not the starting point of man's ethical endeavor. If science progressively reveals one part of human nature, ethics progressively reveals another. It is unreasonable to suppose that science approaches a unique limit called reality while ethics meanders without a goal.

Earlier, the claim was made that the constructive view we hold provides transcultural ethical standards. Yet the flux it envisions and the present diversity of mores appear to contradict this statement. What this means is that the truth of the method is not automatic, that it will not be true unless we *make* it true. For it asks us to join hands with all men in a common constructive effort, where static value doctrines were content to let us behold and speculate. But there is even more solid ground for believing that transcultural standards already exist, albeit in limbo. Sympathetic observation shows convincingly, I believe, that today's functioning ethical systems do not differ markedly at all. There is a far greater common core of imperatives and primary values than an analysis of the middle ground, the tangle of secondary values, is able to unveil: the appearance of diversity arises because the true constructive nature of ethics has been obscure and the emphasis on values unhappily focused attention in the wrong place.

It is true, our treatment has not led us to name specific imperatives or specific primary values. The kind of work which distills these items in their quintessence from the impure substance of our culture has yet to be done. The urgency of this task, which must be completed before intercultural comparisons become meaningful, is clear and present. But the principles that underlie, and in the sense of validation here proposed ought to underlie, our individual behavior are neither obscure nor ambiguous, for the agreement as to personal ethical conduct is global and is amazing.

It makes little difference whether you choose as the source of your imperatives the Sermon on the Mount, the Koran, the Analects of Confucius, the eightfold path of the Buddha or the Tao—none of these conflict within the imperfect precision that characterizes their commands. Nor does it matter greatly whether you strive for happiness, self-fulfillment, universal love or the peace that passeth understanding. But make these commitments work in your life —and do not worry unduly about values.

In the words of Pascal, "All the good maxims have been written. It only remains to put them into practice."

Author Index

Adler, N. J., 136
Ahriman, 239
Aquinas, 195
Aristotle, 33, 91, 151, 162, 163, 212, 239
Auerbach, Felix, 76
Augustine, 195
Aurelius, Marcus, 245
Ayer, A. J., 133, 135, 214

Babbitt, I., 136
Bainton, Ronald, 250
Barth, 136
Bax, E. B., 136
Bentham, J., 204, 207
Bentley, Jeremiah, 204
Berdaev, N., 137
Bergson, H., 137
Blackwell, D. H., 118
Blake, R. M., 136
Blanshard, Brand, 173, 174, 244, 274, 275
Bolyai, 145
Breastead, J. H., 235
Brentano, 214
Brightman, E. S., 137
Broad, C. B., 107, 137
Brooks, F. C., 118
Brunner, E., 136
Butcher, S. H., 161
Butler, J., 271 et seq.

Calkins, Mary W., 137
Cabot, R. C., 137

Carnap, R., 13, 135
Carritt, E. F., 137
Cassidy, H. G., 6
Cassirer, E., 17, 21, 23
Cato, 225
Churchman, C. W., 101
Cicero, 124, 162
Columbus, 141
Comte, A., 71
Confucius, 149, 154, 188, 195, 233, 252, 254, 293
Coulomb, 79, 268

Descartes, 25, 258
Dewey, J., 136
Diaz, 70
Dirac, P.A.M., 196
Dodd, S. C., 89
Drake, Durant, 136
Ducasse, C. J., 180
Durkheim, E., 39, 80, 81, 94, 136

Eddington, 18
Einstein, 7, 22, 43, 145, 196, 230, 231, 282
Engels, 136
Euclid, 40, 144, 145, 146, 171, 202
Everett, W. G., 137
Ewing, A. C., 137

Feigl, H., 135
Fichte, 214

INDEX

Field, G. C., 137
Fite, W., 136
Francis of Assisi, 191
Frank, P., 282
Freud, S., 135, 231, 273, 274

Galileo, 43, 91, 230, 231
Garnett, C. B., 136
Gauss, 51
Gautama, 202, 233
Gibbs, 39
Gilson, 136
Goethe, 65, 168, 249

Haldane, 136
Hall, E. W., 128
Hamilton, 35, 36, 38
Hammurabi, 153
Hare, R. M., 106, 174
Harris, Errol, 143
Hartman, N., 137
Hartman, R. S., 99, 130 et seq., 151, 214
Hegel, E. G., 227
Heidegger, 63
Heinemann, F., 130
Heisenberg, 63, 196, 279
Helmholtz, 259
Hilbert, 33
Hill, T. E., 99, 135
Hitler, 178, 286
Hocking, W. E., 137
Hume, 9
Husserl, 46
Huxley, J., 136

Jacob, 240
Jefferson, 188

Kallen, H., 136
Kant, 9, 16, 21, 83, 144 et seq., 193, 195, 200, 208 et seq., 226, 227, 271, 272, 280
Kautsky, 136
Kelvin, 259
Kepler, 158, 259
Kinsey, 124
Kropotkin, P., 136

Laird, J., 137
Lao Tse, 160, 233, 289
Leibnitz, 25
Levine, Israel, 136
Levy-Bruhl, 136
Lewis, C. I., 136
Lin, Yu Tang, 149
Lindsay, R. B., 6, 83, 144
Lippmann, W., 147
Lobatchevski, 145
Locke, 9, 184, 188, 195, 230
Lotze, 227
Lundberg, G., 75, 89, 179
Luther, 202, 249, 250, 251, 252, 253

MacDougall, W., 136, 215
MacKaye, J., 136
MacKenzie, J. S., 137
Malevich, 90
Mannheim, R., 135
Manu, 202
Maritain, 136
Marx, 114, 136
Maslow, 129, 245
Maxwell, 29
Mead, G. H., 136
Melanchthon, 250
Mendel, 230
Mill, J. S., 204, 205, 206, 207
Minkowski, 145
Mohammed, 202

INDEX

Moore, G. E., 105, 106, 134, 137, 214
Moses, 153, 170, 202
Muirhead, J. H., 137
Munsterberg, H., 137

Newton, I, 25, 31, 43, 65, 79, 126, 158, 258, 264
Niebuhr, R., 136
Nietzsche, 139, 244, 273, 274
Northrop, F. S. C., 14 et seq., 70, 136, 181 et seq.
Nyrgen, A., 243

Occam, 30
Ogden, 135
Osiris, 233, 236, 237, 242
Otto, M. C., 136

Pap, A., 135
Pareto, 135
Parker, D. H., 136
Pascal, 245, 293
Paton, H. J., 137
Paul, 195, 202, 233, 242 et seq., 249, 250, 251, 252
Percy, R. B., 136
Perry, C. M., 135
Pharao, 236
Philodemus, 243
Plato, 38, 82
Poincaré, 21, 38
Prall, D. W., 136
Pratt, J. B., 136
Pritchard, H. A., 137
Protagoras, 282
Pythagoras, 258

Ranke, 6
Rashevski, N., 205

Reid, J. R., 136
Ricardo, 114
Rice, P. B., 106
Richards, 135
Riemann, 145, 146
Rogers, A. K., 135
Ross, W. D., 137, 266
Russell, 135

St. Francis, 202, 283
St. Thomas, 212
Santayana, G., 136
Scheeler, M., 245
Schelling, 214
Schlick, M., 135, 136
Schopenhauer, 214, 245, 273
Schrödinger, 196
Schweitzer, 136, 283
Seneca, 162, 225, 245
Seth, J., 137
Shand, A., 136
Shannon, 73
Sharp, F. C., 135
Sheldon, W. H., 136, 207, 222, 247, 248
Sidgwick, H., 206
Smith, Adam, 114, 214
Smith, N. M. Jr., 108, 118
Smith, T. V., 135
Socrates, 82, 221
Sorley, W. R., 137
Sorokin, P., 42, 80, 82, 213
Spinoza, 82, 195
Stace, W. T., 136
Stapledon, O., 136
Steinbeck, J., 3
Stephenson, 135
Stevens, 214
Stevens, Leslie, 275
Stewart, J. Q., 78
Stoker, R., 272

Sumner, W. G., 135
Sutherland, A., 136
Suzuki, 245

Taylor, A. E., 137
Tennant, F. R., 136
Thamin, M. R., 161
Tillich, 134
Tufts, J., 136

Urban, W. M., 137

Vaihinger, 16
Von Rintelen, F. J., 129, 134

Waddington, C. H., 136
Waley, A., 253

Walters, S. S., 118
Warntz, W., 78
Warren, Earl, 159, 160
Watson, J. B., 135
Werkmeister, W. H., 135
Westermack, E., 135, 205, 206, 214 et seq., 220, 227, 235, 272, 276
Wiener, N., 73
Wittgenstein, 135
Woodbridge, F. E. J., 136
Wright, W. K., 137
Wundt, 218

Zeno, 225
Zipf, G. K., 77 et seq.
Zoroaster, 153

Subject Index

Agape, 243, 246, 247
Altruism, 261, 264
Altruistic love, 213
Analects, 254
Anthropology, 102
Asceticism, 202
Asymptotic realism, 21
Awe, 252
Axiology, 151
Axioms, 144, 150

Beauty, 261, 262
Benevolence, 206, 253
Book of Dead, 237
Brahmanism, 254
Buddha, 153, 254, 289, 293
Buddhism, 253, 254, 286

Carbon dating, 62
Casuistry, 159, 160, 162, 163
Categorical imperative, 200, 208, 228
Causal postulate, 69
Celestial mechanics, 49
Character, 95
Charity, 162, 253
Choice, 110, 171
Christian Ethics, 241
Christianity, 141, 149, 154, 237, 241, 286 et seq.
Codes, 53, 158
Cognitive experience, 11
Cognitivist, 130, 132

Color, 65, 89
Commandments, 148
Commands, 153, 158, 170, 221, 230
Compassion, 248
Competition, 225
Complementarity, 212
Complex ideas, 230
Complexity, 66, 91
Conflict, 266 et seq.
Confucianism, 253
Conscience, 270, 273, 275
Conscientious objector, 268
Control of variables, 68, 91
Coulomb Law, 79, 268
Creative intuition, 259
Cult of Osiris, 236

Data, 12, 45
Death ray, 58
Decalogue, 153, 170
Decisions, 69, 94
Deduction, 41, 44
Deontology, 265
Diversity of mores, 292
Divine Mercy, 249
Duty, 209, 249, 288

Efficacy, 290
Eightfold path, 255
Emotive theory of values, 214
Emotivists, 265
Empiricists, 132
Epicureans, 153

INDEX

Epistemology, 8 et seq.
Equation of state, 68
Ergodic hypothesis, 38
Eros, 243
Error, 278
Est-laws, 289
Esto-laws, 289
Esto-norms, 182, 217
Ethical laboratory, 178
Ethical leadership, 281
Euclidean geometry, 144 et seq.
Evolutionary goals, 222
Evolutionism, 239
Exactness, 290
Expediency, 286
Experience, 9
Explication, 158

Faust, 249
Formalists, 132
Freedom, 96, 285
Free will, 92
Friendship, 264
Frustration, 251
Functions, 87

Gefühlsevidenz, 218
Geisteswissenschaften, 9
Genius, 150
Gewissensangst, 274
Golden Rule, 154, 225
Good, 265 et seq.
Good will, 210
Grace, 203, 241
Group enterprise, 276

Happiness, 167, 232, 258
Hari-kari, 81
Hedonism, 200, 201, 207
Hermits, 203
Heroism, 280

Honesty, 225, 261, 262
Hue, 66
Human life, 175
Humility, 251

Id, 231
Immortality, 237
Imperatives, 151 et seq., 256, 283
Indeterminism, 94
Indicative, 151
Indiscriminate love, 242
Induction, 41, 44, 234, 236
Inductive leap, 42, 224, 231
Inspiration, 42, 150
Instinct, 276
Intangibles, 117
Introspection, 46
Invariance, 87, 282
Islam, 237

Jahwe, 240
Judgments, 217
Judaism, 233, 237, 240
Jurisprudence, 155
Justice, 251, 261, 264
Justification, 251

Kamikaze, 81
Karma, 255

Language, 262
Law, 269
Laws, 157
Liberty, 258
Libido, 6
Libri poenitentiales, 162
Love, 243
Loyalty, 285
Lumen naturale, 144

Marginal utility, 116
Maximum principles, 83

INDEX

Mazdaism, 233, 239
Mercy, 250
Metaphysics, 236
Methodology, 141
Minimum principles, 83
Monotheism, 240
Moral approval, 214
Moral disapproval, 214
Moral emotions, 216
Moral judgment, 174
Moral suasion, 290
Multiplicity of ethical systems, 291

Naturalists, 132
Nirvana, 255
Non-cognitivists, 132
Normative elements, 125
Norms, 158, 288

Objective reference, 221
Obligation, 128, 244, 262, 264, 288
Occam's razor, 30
Oedipus complex, 274
Operational definition, 258
Operators, 87
Opportunism, 286
Ought, 120, 123, 168, 251

Parallelism between Ethics and Science, 260
Patriotism, 285
Paulinism, 242, 244
Physical law, 279
Positivism, 103, 135
Postulates, 30
Postulates of science, 53
Preferences, 124
Price, 113, 121
Primary values, 200, 230
Principles of validation, 53, 209, 256
Principles of verification, 53

Process theories, 136
Propagation of rumors, 79
Protocol, 13, 273
Prudence, 286
Purpose, 288
Pythagorean Theorem, 258

Qualities, 64, 90
Quantities, 90

Rank-size law, 76
Reality, 14
Redemption, 241
Reformation, 162
Refutation, 60
Relativity, 281 et seq., 284
Religion, 236
Remorse, 249
Repentance, 248
Retributive emotions, 215
Riemannian geometry, 146
Right, 265
Rival theories, 287
Rules of correspondence, 28, 264

Saturation, 89
Self defeating predictions, 72
Self-denial, 202
Self-fulfilling predictions, 72
Self-fufillment, 167, 212
Self-realization, 137
Simplicity, 269
Single occurrence, 277
Skepticism, 135
Sklavenmoral, 244
Social physics, 78
Social science, 39, 74
Sociology, 102
Statistical postulates, 52
Statutes, 157
Stoicism, 162

Stoics, 153
Subversion, 285
Suicide, 80
Superego, 274
Supernatural, 131
Supraconscious, 42
Survival, 213
Suttee, 81
Symbolic forms, 23
Symmetry, 87, 90

Tao, 154
Taoism, 253
Temperature, 26, 100
Theory formation, 259
Thermodynamic imperative, 83
Thermodynamics, 37, 259
Tragedy, 267
Transaction, 71
Truth, 261, 262

Uncertainty principle, 63, 85, 279
Universal rule, 227
Upanishads, 254
Utilitarian doctrine, 204

Validation, 169, 171
Validity, 128
Values, 64, 99, 104, 105, 168, 261, 266
Variability, 283
Vedas, 254
Verifacts, 168, 264
Verification, 44
Virtue, 176

Yang, 253
Yin, 253

Zend Avesta, 239